IONIZING RADIATION EFFECTS AND APPLICATIONS

Edited by **Boualem Djezzar**

Ionizing Radiation Effects and Applications
http://dx.doi.org/10.5772/intechopen.68295
Edited by Boualem Djezzar

Contributors

Alexandr Gradoboev, Anastasiia Simonova, Olga Babich, Ksenia Orlova, Tatsuo Shiina, Romain Coulon, Jonathan Dumazert, Mohamed Ghobashy, Betül Çalişkan, Slavica Brkić, Igor Makarov, Alexander Ponomarev, Samy Elsayed

Published by InTech
Janeza Trdine 9, 51000 Rijeka, Croatia

Publishing Process Manager Dajana Pemac
Technical Editor SPi Global
Cover InTech Design team

First published March, 2018
Printed in Croatia
Legal deposit, Croatia: National and University Library in Zagreb

Additional hard copies can be obtained from orders@intechopen.com

Ionizing Radiation Effects and Applications, Edited by Boualem Djezzar
p. cm.
Print ISBN 978-953-51-3953-9
Online ISBN 978-953-51-3954-6

For EU product safety concerns:
IN TECH d.o.o., Prolaz Marije Krucifikse Kozulić 3, 51000 Rijeka, Croatia,
info@intechopen.com or visit our website at intechopen.com.

Meet the editor

Dr. Boualem Djezzar, PhD, received the Diplôme des Etudes Supérieures (DES) in solid-state physics from the U^{ty} of Constantine, Algeria; the Diplôme des Etudes Approfondies (DEA) in microelectronics from U^{ty} of Grenoble, France; the MSc degree in thin film from U^{ty} of Algiers, Algeria; and the PhD degree in microelectronics from U^{ty} of Boumerdès, Algeria. He is currently affiliated with CDTA research center. He is the team leader of the semiconductor component reliability group. His current research interests include CMOS technology reliability, especially electrical characterization and modeling of radiation and NBTI effects. He is the author and coauthor of more than 70 papers published in refereed journals and refereed conference proceedings on characterization, modeling, and simulation of MOS device reliability.

Contents

Preface

Contrary to the common belief, ionizing radiations have beneficial applications in different fields, including environmental, medical, industrial, agricultural, and semiconductor fields. This fact has been achieved thanks to the fundamental understanding of the elementary interactions of radiations with matter. Indeed, several studies and developments have been carried out on radiations since their discovery, generating many research axes and establishing various technologies desired by different industries. This book addresses some aspects of ionizing radiation effects as well as their applications. Through its chapters, the book covers various applications of radiation/matter interactions, hardening process for semiconductor devices, biomedical imaging by radiation luminescent quantum dots (QDs), hydrogen gas detection by Raman lidar sensor for explosion risk assessment, water and wastewater purification by radiation treatment for environment, doping by the neutron transmutation doping (NTD) for semiconductor industry, and polymerization by irradiation, which is useful for industries requiring resistant and protective coating.

The chapters in this book have been written by professors and researchers from academia laboratories and government research centers across the world. This book will be a helpful manuscript for teachers, researchers, postdoctoral research fellows, and senior graduate students from universities and research/development institutions interested in the field of ionizing radiation effects and applications and give focus points for future investigations. The present book consists of eight (8) chapters in a variety of field applications. The general purpose of this book is to provide a comprehensive analysis of ionizing radiations and their applications.

Chapter 1 proposes a hardening method for light-emitting diode (LED) devices using radiation technologies. LED devices, based upon AlGaAs heterostructures, have been submitted to fast neutrons and gamma rays of ^{60}Co source. The initial irradiation by fast neutron particles allows activation of the preexisting defects by decreasing their thermal resistance during the next step tests, especially annealing. As a consequence, the radiation resistance and reliability of LED devices can be enhanced using the combined radiation technologies. **Chapter 2** reexamines the knowledge related to the fluorescent nanocrystal quantum dots (QDs). It investigates their proprieties in terms of size and band gap energy, brightness and photostability, surface coating and water solubility, as well as fluorescence intensity and lifetime. To make them useful for biomedical applications, QDs need to be conjugated to biological molecules without influencing their function. The chapter also describes their functionalization as medical diagnostic tools in order to probe and image the biomolecules and cells. The near-infrared radiation luminescent QDs are promising for biomedical imaging. **Chapter 3** deals with hydrogen gas detection and concentration monitoring using Raman lidars to prevent explosion accidents of hydrogen gas. Contrary to actual hydrogen sensors, the lidar sensors are contactless types, based on Raman-scattered light, and do not alter the concentration measurement. Two lidars are described; the first is named DPSS laser-based compact Raman lidar and the second is LED-based mini-Raman lidar. They are developed for quantitative measurement of hydrogen gas. The current minimal detection limit of hydrogen gas concentration is 1% at the observation range of 0–50m with the accumulation time of 30 seconds and 0–20m with 3.5 min for DPSS laser-based compact Raman lidar and LED-based mini-Raman lidar, respectively. Raman lidar sensor could be one of the potential solutions for remote sensing device for leaked hydrogen gas, especially when the hydrogen will become an energy resource of daily life. **Chapter 4** provides a basic theoretical knowledge and experiment results on nuclear transmutation. The author presents the semiconductor materials doping by neutron transmutation doping (NTD). Particularly, the conversion of n-type germanium doped with arsenic into p-type germanium by fast reactor neutron irradiation. The insulator-metal transition is analyzed based on the dependence of the DC conductivity on the temperature of the conduction activation energies for different irradiation doses.

In **Chapter 5**, radiation treatment of aqueous systems contaminated with organic compounds and corresponding technologies are addressed. This process is a promising method for water and wastewater purification. The following aspects are given in this chapter: major contaminants of water and wastewater and their sources, interrelation between well-established primary radiation-chemical processes, and the main ways of organic contaminant conversions as well as equipment and sources of ionizing radiation used for aqueous system purification. **Chapter 6** reports on ionizing radiation-induced polymerization, giving its advantages such as the purity of the obtained polymer, while in conventional methods, it needs complicated purification steps. It discusses radiolysis of matter by different ionizing radiation sources like UV irradiation, X-rays, electron beam, and gamma-rays. Since different polymers respond differently to radiation, the quantification of their responses in terms of cross-linking and chain scission is given. Details on radiolysis of water as sources of radicals, which are responsible for polymerization reactions, are also given. In addition, the electron paramagnetic resonance (EPR) method, used for the investigation of the radiation damage centers induced by irradiation in the material structure, is presented in **Chapter 7**. In fact, the interaction of ionized radiation with a material induces some changes in its structure. These changes are called the radical or the radiation damage centers. The latter are seen as free radicals in the form of breaking bonds in the structure and as anion or cation radicals in the form of electron exchange. The EPR spectroscopy method is used to investigate such impairments or paramagnetic centers occurring in structures of organic or inorganic materials. Data processing techniques related to radiation monitoring system are described in **Chapter 8**. Different radiation count rate data processing algorithms are presented. Methods based on frequentist inferences are more appropriate for real-time processing, enabling fast decision-making compared to Bayesian inferences, which are more adapted to postprocessing analyses. Moreover, nonlinear smoother provides an accurate estimation with minimized associated variance. The filtering technique is used to compensate measurement and moving source detection.

Finally, the editor gratefully wants to thank all the contributors of this book. Their insightful contribution helped in the successful completion of the manuscript. I am grateful to the IntechOpen publisher for putting me in charge to edit this book. Finally, I would like to thank my wife and kids, Fatima and Youcef, for their support during the preparation of the book.

Dr. Boualem Djezzar, PhD
Team Leader of Semiconductor Device Reliability
Microelectronics and Nanotechnologies Division
Centre de Développement de Technologies Avancées (CDTA)
Algiers, Algeria

Chapter 1

Application of Radiation Technologies for Quality Improvement of LEDs Based upon AlGaAs

Alexandr V. Gradoboev, Anastasiia V. Simonova,
Ksenia N. Orlova and Olga O. Babich

Additional information is available at the end of the chapter

http://dx.doi.org/10.5772/intechopen.72286

Abstract

The investigation results of the radiation resistance and reliability of light-emitting diodes (LEDs) based upon AlGaAs are presented. The radiation model and the reliability model are described for LEDs. Preliminary irradiation by gamma-quanta and fast neutrons makes it possible to improve the radiation resistance and reliability of the LEDs during further operation. Based on the developed models, radiation technologies are proposed, and the use of which allows increasing the service properties of the LEDs. The suggested technologies can be used for other types of semiconductor devices.

Keywords: light-emitting diodes, AlGaAs, gamma-quanta, neutrons, reliability

1. Introduction

Nowadays infrared wavelength range light-emitting diodes (IR-LEDs) are broadly used in various microelectronic devices that operate in space conditions and at nuclear power plants. Therefore, one already needs to know both their reliability and radiation resistance at the developmental stage. Moreover, operation conditions of LEDs require knowledge of their durability and reliability with complex and combined influence of radiation resistance and long-term operation [1]. In this case, we interpret that complex influence as the simultaneous impact of two or more radiation factors. Accordingly, the combined influence is the impact of two or more radiation factors spread out over a period of time. This is due to the fact that complex and/or combined influence of various types of ionizing radiation is always observed in field operating condition of semiconductor devices. Investigations in this field allow us to develop a radiation model of semiconductor device that describes the changes in its criterial parameters depending on the level and conditions of radiation effect and allows predicting its mode of behavior under the influence. In this case, the criterial parameter of a semiconductor device means such parameter, the change in which limits its working life.

In the same way, the reliability model of semiconductor devices is strictly necessary, which allows to predict the change in criterial parameters during operation.

We should note that it is urgent to add the factors of long-term operation to radiation factors in field operating conditions of the semiconductor devices. Therefore, it is necessary to develop an exploitable model of semiconductor devices [2], which allows describing the change in reliability indexes during long-term operation under conditions of complex and combined influence of various types of ionizing radiation.

At present time, there is an insignificant amount of work on combined influence of ionizing irradiation of semiconductor devices [1, 3–5]. However, works on the combined and complex influence of long-term operation factors and ionizing irradiation are almost completely absent.

Of all the variety of types of ionizing radiation, the research of the influence of irradiation by fast neutrons and gamma-quanta on the changes in the characteristics of various types of semiconductor devices is of primary concern.

Why are these two types of ionizing radiation of special interest? We note that the main effect of neutrons in semiconductor materials is to produce displacement damage. This later can cause shifts in the spectral response, the threshold current, the quantum yield, and the slope efficiency of the LEDs. At that time, ionization mechanism of defect formation dominates for gamma-quanta [6–9]. All other types of ionizing radiation can be represented by means of certain combination of interaction mechanisms of ionizing radiation in terms of their effect on materials. Therefore, the research results of irradiation by fast neutrons and gamma-quanta of various semiconductor device types can be used as a basis for analyzing the influence of other ionizing radiation types and, accordingly, can be used as a basis for developing radiation model of the semiconductor devices.

At the present time, there are no effective and sufficiently reliable methods for calculating the resistance of semiconductor devices to the influence of ionizing irradiation. Therefore, various different simulated equipments are used to determine their resistance to ionizing irradiation [10, 11].

In addition, accelerated tests are generally used to determine reliability indexes of semiconductor devices. They force the aging processes and reduce the length of time required for obtaining the reliability information [12–14]. Generally, the accelerating factors arise from temperature and exaggeration of conditions.

Analysis of the available literature data allows us to conclude that all currently available methods of semiconductor devices' reliability assessment are time-consuming, requiring significant financial expenditures and special equipments. Therefore, obtaining information about reliability and radiation resistance is difficult at the developmental stage of devices.

On the other hand, it is known that ionizing irradiation allows changing directly the parameters of various types of semiconductor devices [1].

The purpose of this work is to develop the radiation technologies focused on improving the reliability of the LEDs based upon AlGaAs heterostructures. The irradiation by fast neutrons and gamma-quanta ^{60}Co is used as a basis of these radiation technologies.

2. Materials and methods

The objects of this investigation were industrial LEDs manufactured on the basis of dual AlGaAs heterostructures with about 2 μm active layer grown on the monocrystalline n^+-GaAs wafer by means of liquid epitaxy. The crystal size was 450 × 450 μm². The investigated LEDs have wide application. Ohmic contact to n^+-GaAs has been made on the (Au-Ge-Ni) basis and for AlGaAs layer on the (Au-Zn) basis.

Double heterostructures significantly exceed single heterostructures and bulk materials in terms of their operation parameters. In particular, the n- and p-layers are made from wide bandgap materials except the absorption. Therefore, the wide-gap window effect is observed [15]. In addition, the used n- and p-layers can be heavily doped. Moreover, the injected electrons and holes are in a very narrow active layer, where n · p is extremely high. It increases the rate of radiative recombination [16]. In addition, LEDs based upon double AlGaAs heterostructures have a higher resistance to ionizing radiation [17, 18]. The above advantages are the main reasons for choosing the object of research.

Figure 1 represents the structure of the double AlGaAs heterostructure of the LED. In the left part of **Figure 1**, the layers of semiconductor materials are shown (the doping impurity is indicated in parentheses). Furthermore, the thickness of the layers is given in the central part of **Figure 1**. Moreover, the distribution profile of the concentration of the main charge carriers and the Al content in the layers are shown on the right part of **Figure 1**.

LEDs were manufactured using standard sandwich technology that involves metallic layer deposition and shaping processes for ohmic contact creation, photolithographic and chemical etching processes for die formation, and dicing for wafer division into individual chips. LEDs had packages and lenses made of an optical compound that was used to form the required angular pattern for output lumen. We emphasize that the preliminary investigation shows us that optical compound irradiation by fast neutrons, and gamma-quanta do not lead to changing its optical properties in given wavelength range. Consequently, we suggest that the changing of the observable light characteristics of the LEDs is due to the changing of characteristics in their active layer only.

In continuous power mode, the LED forward operating current was I_{op1} = 50 mA, and supply voltage U_{op} was not over U_{op} = 2.0 V. The maximum emissive wavelength was within the range of 0.82–0.90 μm. Emissive power at the given operating current was criterial parameter of the LEDs that determined their efficiency.

For every LED emissive power at operating current 50 mA under normal conditions was taken using a measurement complex with spherical photometric integrator. The error did not exceed 5% of emissive power measurement of LEDs. The spread of the emissive power of the initial LEDs for each batch did not exceed ±10%. Moreover, the spread of the operating voltage was ±3%.

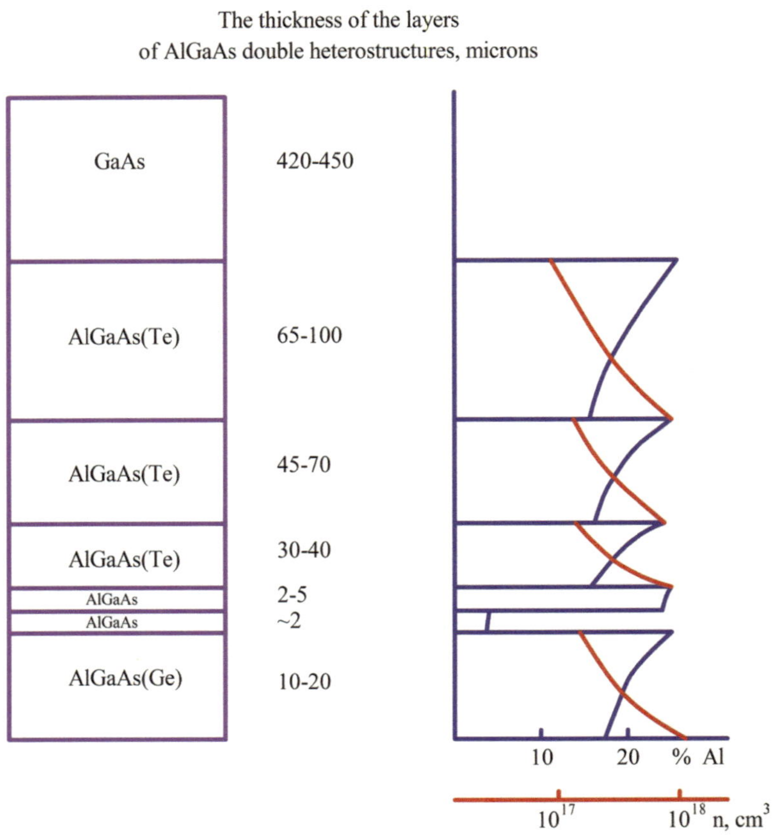

Figure 1. Structure of the double AlGaAs heterostructure.

The named characteristic of LEDs was obtained at the very beginning and after every stage of the investigation. The results were processed by means of mathematical statistics methods. Every batch of LEDs under investigation was characterized by average values of measured parameters.

Irradiation by gamma-quanta was performed with isotopic continuous source of ^{60}Co. The exposure level was characterized by the absorbed dose D_γ [Gy]. The dose rate was about 1 Gy/s, and the average gamma-quanta energy was 1.25 MeV.

Irradiation by fast neutrons was carried out on the pulse simulator installation "Bars-4," which has got a reactor core of the metallic uranium and molybdenum alloy (10% by weight) with a pulse duration of 60 μs. The average neutron energy is 1.4 MeV [19, 20]. The exposure level was characterized by neutron fluence F_n [n/cm^2].

Irradiation by gamma-quanta and fast neutrons was realized in passive power mode, i.e., without the adding external electric field. Moreover, electrical lead of the LEDs was not allowing the flow of electricity.

Long-term operation was modeled using step-by-step tests. Standard certified equipment was used for experiments, and baseplate temperature was 65°C; the increment step of current was $\Delta I = +50$ mA. Operating current of the first stage was $I_{op1} = 50$ mA. Duration of each stage was $t = 24$ h. Every stage of testing was characterized by operating current I_{stepi}. Moreover, each stage of the experiments was distinguished by the temperature of the LED active layer T_{stepi}, which depends on ambient temperature (i.e., baseplate temperature) T_0, LEDs thermal resistance R_T, and consumed electric power on i step U_{stepi}. Therefore, temperature of the LED active layer on each stage of the tests is determined by the following equation:

$$T_{stepi} = T_0 + R_T + I_{stepi} \times U_{stepi}. \tag{1}$$

Therefore, we used both temperature and operating current as acceleration factors during step-by-step tests. Furthermore, limit step of the tests did not go beyond catastrophic failure (CF) development [21, 22]. The step-by-step tests were stopped when at least 80% of the LEDs were out of service in each batch of the research LEDs.

The following batches of the LEDs were composed for the research. The quantity of the LEDs in each batch was 20 items:

- LED-1 and LED-2 were for resistance research of irradiation by gamma-quanta and fast neutrons.
- LED-3 was for research of reliability.
- LED-4 and LED-5 were for research of influence of preliminary irradiation by gamma-quanta with further annealing on reliability.
- LED-6 and LED-7 were for research of influence of preliminary irradiation by fast neutrons with further annealing on reliability.

Rationale selection of preliminary irradiation level by gamma-quanta and fast neutrons, and annealing parameters will be considered below.

3. Radiation resistance and reliability of the LEDs

3.1. Radiation resistance of the LEDs

Consider the results of a study of the resistance of batch LED-1 to ionizing irradiation. **Figure 2** shows the relative change of emissive power measured at the operating current depending on the absorbed dose of gamma-quanta [23]. Here and further, emissive power P measured after exposure is normalized to its initial value P_0 of LEDs. The absorbed doses of $D_{\gamma 1}$ and $D_{\gamma 2}$ in **Figure 2** are explained below in the text.

Whereas **Figure 3** depicts the relative change of emissive power of the batch LED-2 measured at the operating current depending on fluence of fast neutrons [24]. The fluences F_{n1} and F_{n2} in **Figure 3** are explained below in the text.

We note that the spread of emissive power of LEDs in the batch rises to ±15% of average value in the batch when dose of irradiation by gamma-quanta and fluence of fast neutrons increase. Accordingly, the spread of operating voltage of the LEDs rises to ±5%.

These research results suggest the following radiation model of the LEDs. It describes the changes of emissive power of the LEDs under irradiation by gamma-quanta and fast neutrons:

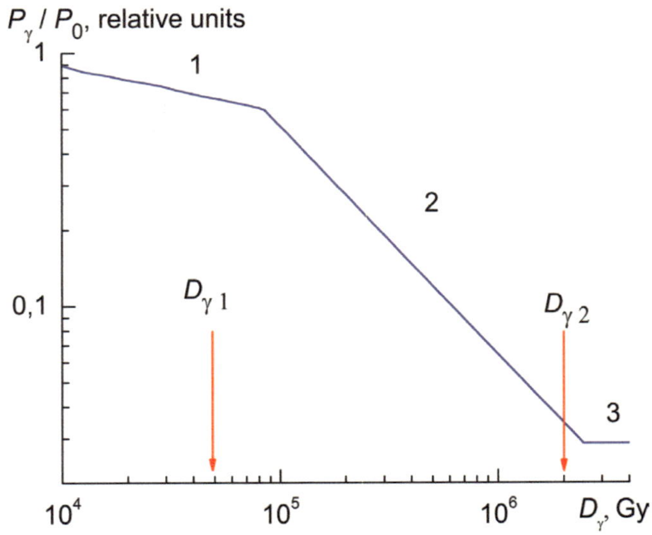

Figure 2. Relative decrease of the emissive power of the LEDs depending on gamma-quanta irradiation dose: 1, 2, and 3, stages of the emissive power decrease; $D_{\gamma 1} = 5 \cdot 10^4$ Gy; $D_{\gamma 2} = 2 \cdot 10^6$Gy.

- In the first stage, the fall of emissive power of the LEDs is attributed to the radiation-stimulated reconstruction of the initial defect structure of the LED crystal (field 1 in **Figures 2** and **3**) as evidenced by saturation of this stage as the level of exposure increases.
- In the second stage, the fall of emissive power of the LEDs during irradiation is attributed solely to the introduction of radiation defects (field 2 in **Figures 2** and **3**).
- In the third stage, the LED transits into the field of low electron injection into the active layer of the LED (field 3 in **Figures 2** and **3**).

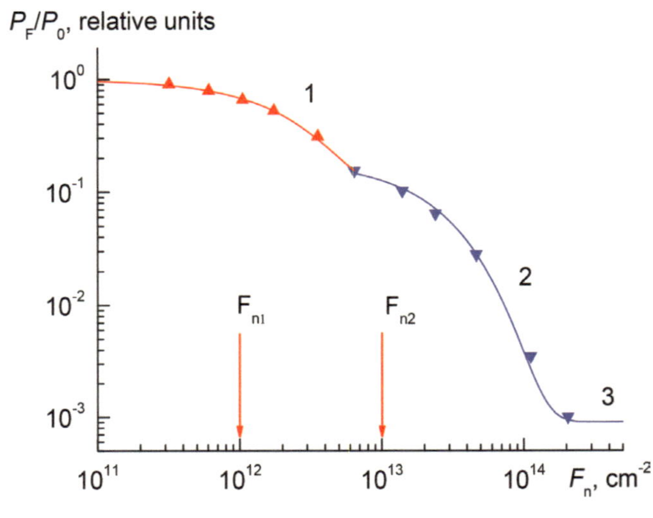

Figure 3. Relative decrease of the emissive power of the LEDs depending on fast neutron fluence: 1, 2, and 3, stages of the emissive power decrease; $F_{n1} = 10^{12}$ n/cm^2; $F_{n2} = 10^{13}$ n/cm^2.

We note that the low electron injection mode is characterized by weak dependence of emissive power of the LEDs on operating current value [25].

It should be particularly emphasized that in the third stage, the CF is observed and caused by accelerated degradation of ohmic contacts' "metal semiconductor" [22, 26].

Additionally, the type of the ionizing radiation determines the relative contribution of the first stage of the emissive power decrease of the LEDs, which is due to the influence of the type of ionizing radiation on the efficiency of the reconstruction of the initial defect structure. We compared the experimental results presented in **Figures 2** and **3**. Therefore, under irradiation by gamma-quanta, the contribution of the first stage of the emissive power decrease of LEDs is about 40% of the initial emissive power, while for fast neutrons, it is about 85%.

This radiation model of the LEDs completely corresponds to previously discussed model of LEDs based on other materials [27–29]. Therefore, the radiation model of decrease of emissive power of the LEDs can be extended to practically all LEDs made of various semiconductor materials.

Each of the named distinctive stages of the emissive power fall of the LEDs under irradiation can be described by the corresponding empirical relationships with their own damage constants. They are usually used to describe the processes of changing the criterial parameters of devices under various influences [30, 31]. The established relationships are capable of predicting the radiation resistance of the LEDs.

3.2. Reliability of the LEDs

Next, consider the changes of emissive power during operation and more specifically during step-by-step tests. **Figure 4** presents relative change of emissive power of the batch LED-3 during step-by-step tests.

Decrease of emissive power during step-by-step tests can be characterized by three distinctive stages, which we observed previously while investigating the radiation resistance of the LEDs (part 3.1, **Figures 2** and **3**). One might assume that the first stage of step-by-step tests decrease in LED emissive power (field 1, **Figure 4**) is possible due to rearrangement of original defect structure exposed to long-term operation factors. Accordingly, emissive power gets reduced on the second stage as the result of inducing new structural defects under influence of long-term operation factors (field 2, **Figure 4**). On the third stage of emissive power fall during long-term operation (field 3, **Figure 4**), we can observe a transition into the mode of low electron injection with further origination of CF. We note that degradation of ohmic contacts' "metal semiconductor" is a preliminary to CF during step-by-step tests [22, 26].

The identity of determined stages of the emissive power decrease of the LEDs under influence of ionizing radiation and long-term operation factors allows for the conclusion that the long-term operation factors correspond to the radiation factors according to their physical nature. In this case, one may talk of one whole model of the emissive power degradation of the LEDs due to the influence of ionizing radiation and long-term operation factors. At the same time, it should be specially noted that proper correlations between the established stages are observed for each of these factors. In actual fact, the contribution of

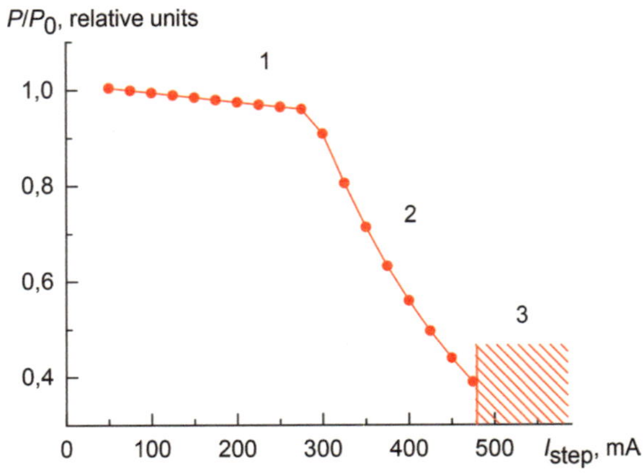

Figure 4. Relative change of the emissive power of the LEDs depending on step number of the tests: 1, 2, and 3, stages of the emissive power decrease.

each determined stages to the overall decrease of the emissive power depends on type of influencing factors. In addition, each stage is characterized by its own value of the damage constants for each type of influencing factors.

The above-presented results have proved the identity of degradation processes in LEDs under the influence of ionizing radiation and factors of long operating time. Therefore, it has been established that under the influence of ionizing radiation on the LED and long-term operation factors, the similar stages of the radiation power decrease are observed. In this case, the first stage in both cases is characterized by saturation, which made it possible to associate its appearance with rearrangement of the initial structure of the defects. Furthermore, at the first stage, the emissive power decrease is due to identical defects and is independent of their appearance history. Similar reasoning is applicable to other determined stages. We note that the relative contribution of the stages to the overall emissive power decrease of the LEDs is defined by the type of influence.

The above-described radiation and reliability models of LEDs can be used as a basis for the development of the exploitable model of LEDs. It is a complex of relationships that describe the change in the emissive power of LEDs (criterial parameter) under the complex and combined influence of various types of ionizing radiation and long-term operation factors. The practical application of the exploitable model of LEDs will make it possible to predict the behavior of LEDs under different operating conditions. The proposed model is universal, because it describes the change in the emissive power of LEDs under the influence of damaging factors of different nature.

The physical nature of the exploitable model of the LEDs allows using it as a base for the development of exploitable models of other types of semiconductor devices.

These research results make it possible to recommend the radiation technology for improving the operational parameters of the LEDs [32, 33].

4. Radiation technologies for improving reliability of the LEDs

4.1. Preliminary irradiation by gamma-quanta

Consider the influence of preliminary irradiation by gamma-quanta with further annealing of radiation-induced defects over change of emissive power during step-by-step tests.

Doses $D_{\gamma 1}$ for batch LED-4 and $D_{\gamma 2}$ for batch LED-5 were used for preliminary irradiation by gamma-quanta. The irradiation dose $D_{\gamma 1}$ corresponds to the first stage of decreasing the radiation power of the LEDs irradiated by gamma-quanta. Moreover, the dose $D_{\gamma 2}$ corresponds to the second stage of the emissive power fall of the LEDs (see **Figure 2**).

After preliminary irradiation by gamma-quanta, the annealing was carried out. Moreover, the annealing mode corresponds to the first stage of the step-by-step tests described above. **Table 1** represents the change of the emissive power after preliminary irradiation with further annealing and during step-by-step tests. Furthermore, almost complete recovery of the emissive power of the LEDs was observed to their initial values after preliminary irradiation with annealing. The measurement results were used as initial values for further research of reliability.

The use of selected preliminary irradiation doses makes it possible to obtain information about the correlation between the processes of decreasing the emissive power of the LEDs caused by the combined influence of ionizing radiation and long-term operation factors.

In the results of step-by-step tests, the LED-4 batch is very neatly divided into two distinctive subgroups LED-4a and LED-4b when the step of the tests rises.

Next, we consider obtained results in more details. **Figure 5** depicts the change of the emissive power of the LEDs from subgroup LED-4a during step-by-step tests. Furthermore, the subgroup LED-4a is divided into two subgroups LED-4a1 (20% from the batch LED-4) and LED-4a2 (35% from the batch LED-4). Moreover, **Figure 5** illustrates for comparison the change of the emissive power of the LEDs without preliminary irradiation (LED-3 batch) during operation.

Research stage	**Level of emissive power, % from initial value**							
	LED-3	**LED-4**				**LED-5**		
		LED-4a		**LED-4b**		**a**	**b**	**c**
		a1	**a2**	**b1**	**b2**			
After preliminary irradiation by gamma-quanta	–	95				8		
After annealing	–	102				97		
Initial values before step-by-step tests	100	100				100		
Before CF development	39	82	67	65	56	107	136	85
Step of the CF appearance (I_{stepi}, mA)	475	550	500	500	450	400	400	450

Table 1. The emissive power changes of the LEDs on the different stages of radiation technology implementation.

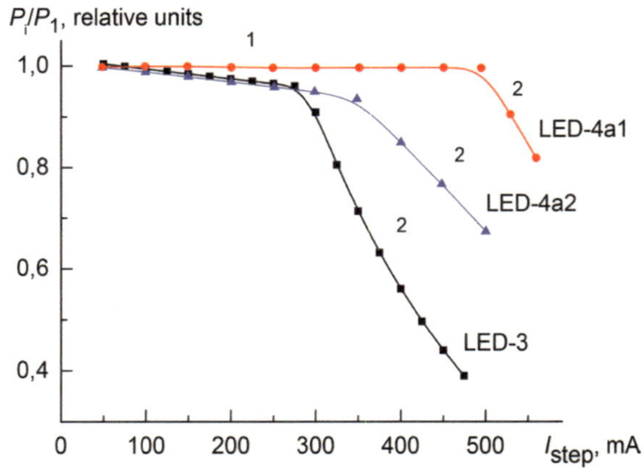

Figure 5. Relative change of emissive power of subgroup LED-4a during step-by-step tests: 1 and 2, revealed stages of the emissive power fall; LED-4a1 and LED-4a2, marked subgroups; LED-3, batch LED-3.

It is apparent that all of the presented dependences have the same form. However, the dependences differ according to the contribution of the first stage in total decrease of the emissive power of the LEDs during step-by-step tests. Therefore, **Figure 5** shows that preliminary irradiation by gamma-quanta makes it possible to decrease the contribution of the first stage (up to its complete elimination for the subgroup LED-4a1) to a total decrease of the emissive power during step-by-step tests.

Each of the marked subgroups can be characterized by the eigenvalues of the efficiency coefficient η for the LEDs after preliminary irradiation and further annealing. In this case, the efficiency coefficient means the ratio of the emission power of the LEDs to the power consumption. Here and elsewhere, the efficiency coefficient is average value for the batches and the marked subgroups. Furthermore, the change of efficiency coefficient can be described by the following equation when passing from one subgroup to other subgroups:

$$\eta_{\text{LED-3}} < \eta_{\text{LED-4a2}} < \eta_{\text{LED-4a1}} \tag{2}$$

Moreover, the dependence between emissive power fall and efficiency is absent in an explicit form for LED-3 batch.

Figure 6 represents the established dependence of the emissive power fall during step-by-step test for subgroup LED-4b. It is divided into two subgroups LED-4b1 (35% from the batch LED-4) and LED-4b2 (10% from the batch LED-4).

It can be seen that the subgroup LED-4b is divided into two subgroups, in which the change of the emissive power is determined by the corresponding value of the efficiency by the following equation:

$$\eta_{\text{LED-4b2}} < \eta_{\text{LED-4b1}} \tag{3}$$

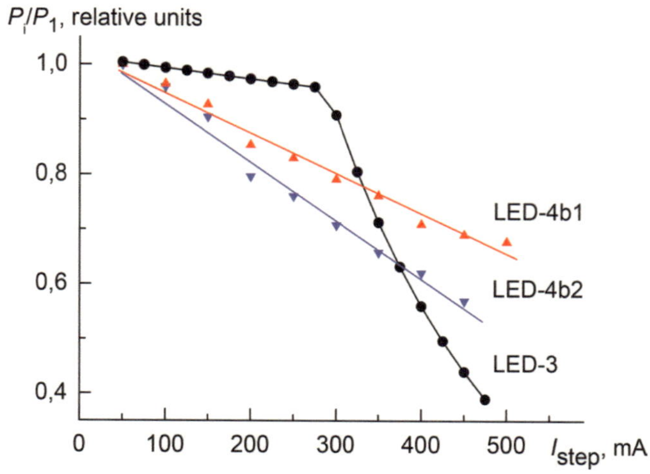

Figure 6. Relative change of the emissive power during step-by-step tests for subgroup LED-4b: LED-4b1; LED-4b2, marked subgroups; LED-3, batch LED-3.

Consequently, we obtain the following equation taking into account Eq. (3):

$$\eta_{LED\text{–}4b2} < \eta_{LED\text{–}4b1} < \eta_{LED\text{–}3} < \eta_{LED\text{–}4a2} < \eta_{LED\text{–}4a1} \quad (4)$$

Certainly, the additional research is necessary to identify the parameters that determined their specific difference for the marked subgroups of the LEDs.

In addition, the preliminary irradiation by gamma-quanta with the dose corresponding to radiation-stimulated rearrangement of the initial defect structure makes it possible to increase the resistance of ohmic contacts to the influence of long-term operation factors. Furthermore, it allows decreasing the probability of the CF development and increasing the reliability of LEDs.

Therefore, the possibility of practical application is shown to except almost completely the contribution of the first stage of decreasing the emissive power of the LEDs during step-by-step tests. It is realized by preliminary irradiation by gamma-quanta with a dose $D_{\gamma1}$ and further annealing. This technology allows improving significantly the operational characteristics of the LEDs. The use of different levels of preliminary irradiation by gamma-quanta within the first stage (see **Figure 2**) allows controlling these processes. Furthermore, it makes possible to control the contribution of the first stage of emissive power decrease of the LEDs during operation.

Moreover, the used doses of preliminary irradiation by gamma-quanta, annealing temperature, and annealing duration are not optimal. Consequently, when the proposed radiation technology is optimized, the obtained results can exceed the values presented here.

Therefore, the preliminary irradiation by gamma-quanta with dose corresponding to the first stage of emissive power decrease of the LEDs (see **Figure 2**) with further annealing makes it possible to increase the reliability of the LEDs.

Now, turn to consider the results of step-by-step tests for batch LED-5. The LEDs of this batch were preliminary irradiated by gamma-quanta with the second dose corresponded to the second stage of the emissive power fall under irradiation by gamma-quanta (**Figure 2** and **Table 1**). We emphasize that the emissive power decrease during step-by-step tests for batch LED-5 fundamentally differs from the above-observed emissive power fall for batch LED-4. The emissive power of LEDs from the batch LED-5 (similar to batch LED-4) almost completely returns to the initial value of the emissive power after the annealing, in spite of the fact that emissive power of the LEDs decreased significantly after preliminary irradiation by gamma-quanta. This is shown in **Table 1**.

Figure 7 depicts the change of the emissive power during step-by-step tests for the batch LED-5. The batch is divided into three equal in percentage ratio subgroups (LED-5a, LED-5b, LED-5c in **Figure 7**). It should be specially noted that the observed changes of emissive power from the LED-8 batch practically do not lead to a LED failure as a result of its decrease. Moreover, the reliability of the LEDs is limited only by the development of CFs.

Two strongly marked peaks of the emissive power increase during step-by-step tests are characterized in all marked subgroups. Furthermore, some of the detected peaks clearly have a compound shape.

Each of the marked subgroups can be characterized by the eigenvalues of the efficiency coefficient for the LEDs after preliminary irradiation and further annealing. Furthermore, the change of efficiency coefficient can be described by the following equation when passing from one subgroup to other subgroups:

$$\eta_{\mathrm{LED\text{-}5c}} < \eta_{\mathrm{LED\text{-}5b}} < \eta_{\mathrm{LED\text{-}5a}} \tag{5}$$

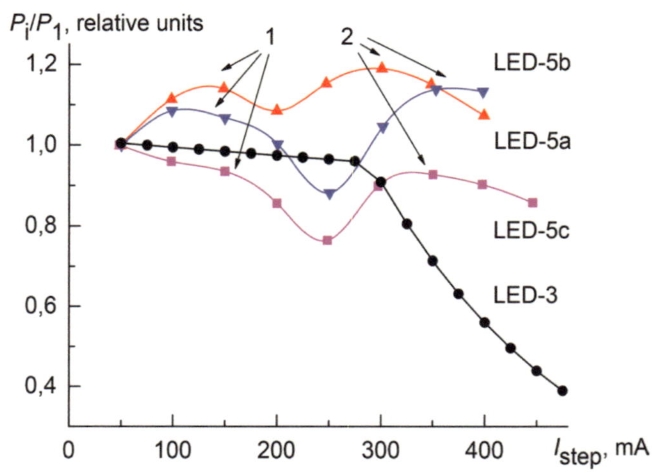

Figure 7. Relative change of the emissive power during step-by-step tests for batch LED-5: 1 and 2, recovery of the emissive power; LED-5a, LED-5b, and LED-5c, marked subgroups; LED-3, batch LED-3.

Comparing the results of the step-by-step tests for batches LED-5 and LED-3, it allows concluding that the observable annealing is probably attributed to annealing radiation defects in particular (possibly stimulated by an external electric field). The defects' precipitation is typical for the second stage of the emissive power fall of the LEDs during step-by-step tests.

The effective annealing temperatures can be estimated to the maximum of the emissive power rise using Eq. (1) for marked subgroups of batch LED-5. Therefore, the annealing of two distinctive types of defects is observed for each of marked subgroups of LED-5 batch. Moreover, the effective annealing temperature is in the range of 348–352 K for the first defect. Furthermore, the effective annealing temperature is in the range of 362–370 K for the second defect. Certainly, the effective temperatures have rough estimate that obtain in such a way. At present time, we cannot connect annealing defects with concrete types. This is subject for further research.

Analysis of the results presented in **Table 1** allows concluding that the preliminary irradiation by gamma-quanta with dose $D_{\gamma 2}$ leads to earlier appearance of CF. It is due to accelerated degradation of ohmic contacts.

Preliminary irradiation by gamma-quanta in the field of radiation defects' introduction leads to accelerated degradation of ohmic contacts and increasing the probability of CF development. Consequently, it leads to decrease the ultimate reliability of the LEDs.

The obtained results allow us to recommend the preliminary irradiation by gamma-quanta in the manufacturing technology of the LEDs to improve their operational parameters.

4.2. Preliminary irradiation by fast neutrons

Consider the influence of preliminary irradiation by fast neutrons with further annealing the radiation-induced defects over change of the emissive power of the LEDs during operation.

The change in the emissive power of the LEDs after preliminary irradiation and further annealing is presented in **Table 2**. Here, the values of the emissive power of the LEDs measured before the CF development are shown. A partial or complete recovery of the emissive power of the LEDs to the initial values is observed after annealing. These obtained results were used as initial values for further research of the reliability.

For batch LED-6, the preliminary irradiation by fast neutrons was chosen in the field of the first stage of the emissive power fall (**Figure 3**). **Figure 8** represents the change of the emissive power for batch LED-6 during step-by-step tests. Moreover, **Figure 8** illustrates the emissive power change for LED-3 batch in contrast. Furthermore, the batch LED-6 is divided into two distinctive subgroups LED-6a (55% from the batch LED-6) and LED-6b (45% from the batch LED-6).

The following equation of the efficiency can be composed for marked subgroups of the LED-9 batch:

$$\eta_{\mathrm{LED-6b}} < \eta_{\mathrm{LED-3}} < \eta_{\mathrm{LED-6a}} \tag{6}$$

Research stage	Level of emissive power, % from initial value				
	LED-3	LED-6		LED-7	
		a	b	a	b
After preliminary irradiation by fast neutrons	—	95		14	
After annealing	95	111		32	
Initial values before step-by-step tests	100	100		100	
Before CF development	39	88	73	353	284
Step of the CF appearance (I_{stepi}, mA)	475	500		425	

Table 2. The emissive power changes of the LEDs on the different stages of radiation technology implementation.

Additionally, there are three distinctive peaks of the emissive power recovery for marked subgroups of the batch LED-6. The subgroups differ only in value of the emissive power recovery. The quantities of the peaks show the annealing of three types of defects. Therefore, the observable recovery of emissive power of the LEDs during operation is due to annealing of three types of the defects created by radiation-stimulated reconstruction of the initial defect structure that is stimulated by operation factors.

The effective annealing temperatures can be estimated to the maximum of the emissive power rise using Eq. (1) for marked subgroups of batch LED-6. Therefore, the annealing of three distinctive types of defects is observed for each of marked subgroups of batch LED-6. Moreover, the effective annealing temperature is in the range of 341–345 K for the first defect. Furthermore, the effective annealing temperature is in the range of 355–359 K for the second defect. Additionally, the effective annealing temperature is in the range of 369–376 K for the third defect. Certainly, the

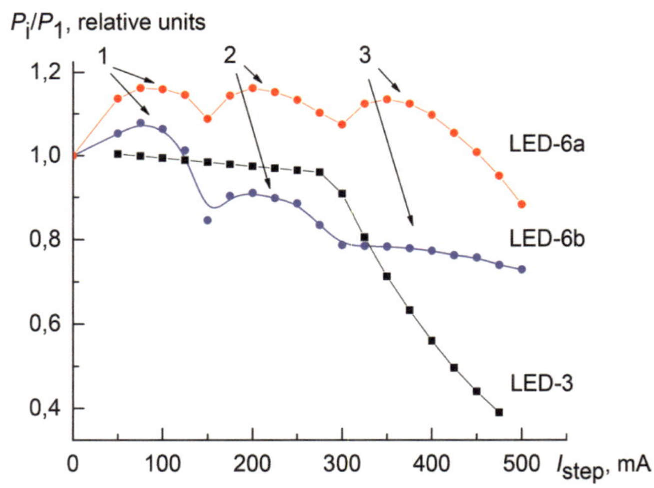

Figure 8. Relative change of the emissive power of the batch LED-6 during step-by-step tests: 1, 2, and 3, the emissive power recovery with further fall; LED-6a and LED-6b, marked subgroups; LED-3, the batch LED-3.

effective temperatures have rough estimate that obtain in such a way. At present time, we cannot connect annealing defects with concrete types. This is subject for further research. Moreover, the reliability increases for the batch LED-6 in comparison with batch LED-3.

Next, **Figure 9** shows the emissive power change of the batch LED-7 during step-by-step tests. This batch is divided into two subgroups LED-7a (70% from the batch LED-7) and LED-7b (30% from the batch LED-7).

For the marked subgroups of the LEDs from the batch LED-7, the following equation can be written as

$$\eta_{LED-7b} < \eta_{LED-7a} \tag{7}$$

There are two stages of the emissive power recovery with its further fall for subgroup LED-7a. This is due to annealing the corresponding defects. Moreover, the rate of the emissive power recovery exceeds significantly the previously observed values for batch LED-6. Therefore, the recovery of the emissive power for batch LED-7 during operation can be considered due to annealing of two types of defects. Additionally, the first of them can be attributed to the field of radiation-stimulated reconstruction of the initial defect structure under influence of the operation factors. The subgroup LED-7b is characterized by the fact that the first stage of the emissive power recovery is absent. The observed changes of emissive power are described by the measurement error. Therefore, the recovery of the emissive power for subgroup LED-7b during operation can be considered due to the annealing of the second defect only.

The effective annealing temperatures can be estimated to the maximum of the emissive power rise using Eq. (1) for marked subgroups of batch LED-7:

- Subgroup LED-7a, $T_{A1} = (360 \pm 2)$ K; $T_{A2} = (388 \pm 2)$ K.
- Subgroup LED-7b, $T_{B2} = (378 \pm 2)$ K.

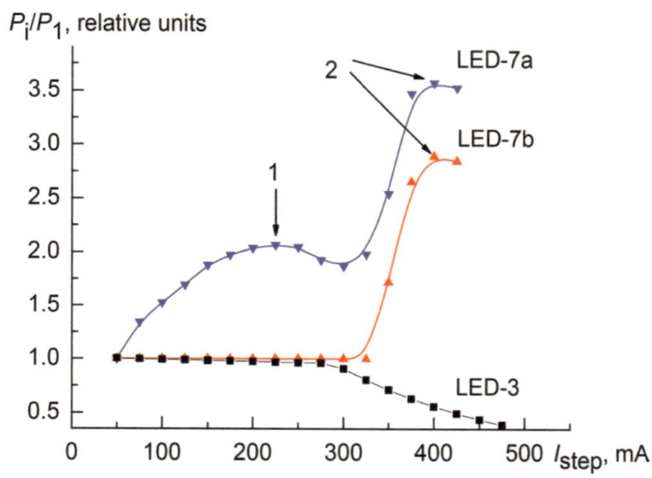

Figure 9. Relative change of the emissive power of the batch LED-7 during step-by-step tests: 1, 2, and 3, the emissive power recovery with further fall; LED-7a and LED-7b, marked subgroups; LED-3, the batch LED-3.

Consequently, the division of the initial LEDs into distinctive subgroups has been established while investigating the influence of preliminary irradiation by fast neutrons with further annealing on the operational characteristics of the LEDs. Each subgroup has its own characteristic dependence of the emissive power decrease of the LEDs during operation.

Furthermore, each subgroup can be characterized by eigenvalues of the efficiency coefficient η for the initial LEDs (after preliminary irradiation with further annealing).

Therefore, preliminary irradiation by fast neutrons makes it possible to activate the initial defect structure by decreasing its thermal resistance during further step-by-step tests.

The obtained results allow us to recommend the present radiation technology for manufacturing the LEDs. This technology is based on preliminary irradiation by fast neutrons and further annealing for improving the operational parameters of the LEDs.

5. Conclusion

The following list sums up the main investigation results of experimental studies described above:

1. The research results of the emissive power change of the LEDs based upon AlGaAs heterostructures under irradiation by gamma-quanta ^{60}Co and fast neutrons. Based on the analysis of the present results, the radiation model of the LEDs has been developed. The established laws can be used to predict the radiation resistance of the LEDs.
2. The investigation of the emissive power change of the LEDs based upon AlGaAs heterostructures during operation based on the step-by-step tests has been presented. Based on the analysis of the research, the reliability model of the LEDs has been developed. The established laws can be used to predict the reliability of the LEDs.
3. Combination of radiation model and reliability model allows to develop an exploitable model of the LEDs. This model makes possible the prediction of the change in emissive power of the LEDs under complex and combined influence of various types of the ionizing irradiation and the factors of long-term operation.
4. Radiation technology for manufacturing LEDs based upon AlGaAs heterostructures with increased radiation resistance and reliability has been presented. It bases on preliminary irradiation by gamma-quanta ^{60}Co and further annealing.
5. Radiation technology for manufacturing LEDs based upon AlGaAs heterostructures with increased radiation resistance and reliability has been developed. It depends on preliminary irradiation by fast neutrons and further annealing.
6. Decision of the optimum conditions of irradiation and annealing allows to improve significantly the radiation resistance and reliability of the LEDs based upon AlGaAs heterostructures, i.e., significantly improve their operation parameters.
7. Suggested complex of the radiation technologies can be recommended for other types of semiconductor devices.

Acknowledgements

This work was supported by the Ministry of Education and Science of the Russian Federation [grant number 03.G25.31.0224 from 03.03.2017].

Author details

Alexandr V. Gradoboev[1*], Anastasiia V. Simonova[1], Ksenia N. Orlova[1] and Olga O. Babich[2]

*Address all correspondence to: gradoboev1@mail.ru

1 National Research Tomsk Polytechnic University, Tomsk, Russia

2 Kemerovo State University, Kemerovo, Russia

References

[1] Gradoboev AV, Surzhikov AP. The Radiation Resistance Microwave Devices Based on Gallium Arsenide. Tomsk: Tomsk Polytechnic University; 2005. 277 p

[2] Gradoboev AV, Simonova AV. Operational model of the electronic products. In: Semipalatinsk Test Site. Radiation Legacy and Development Prospects; September 21-23, 2016; Kurchatov, VKO, Republic of Kazakhstan. Pavlodar: Publishing House; 2016. p. 131-132

[3] Martin DI, Ighigeanu DI, Mateescu EN, Craciun GD, Calinescu II, Iovu HM, et al. Combined microwave and accelerated electron beam irradiation facilities for applied physics and chemistry. IEEE Transactions on Industry Applications. 2004;**40**(1):41-52. DOI: 10.1109/TIA.2003.821655

[4] Busatto G, De Luca V, Iannuzzo F, Sanseverino A, Velardi F. Single-event effects in power MOSFETs during heavy ion irradiations performed after gamma-ray degradation. IEEE Transactions on Nuclear Science. 2013;**60**(5):3793-3801. DOI: 10.1109/TNS.2013.2278038

[5] Liu C, Li X, Yang J, Ma G, Sun Z. Radiation defects and annealing study on PNP bipolar junction transistors irradiated by 3-MeV protons. IEEE Transactions on Nuclear Science. 2015;**62**(6):3381-3386. DOI: 10.1109/TNS.2015.2498201

[6] Chaffin RJ. Microwave Semiconductor Devices: Fundamentals and Radiation Effects. New York: John Wiley and Sons, Inc.; 1973. 387 p

[7] Lang DV, Petroff PM, Logan RA, Johnston Jr WD. Recombination-enhanced interactions between point defects and dislocation climb in semiconductors. Physical Review Letters. 1979;**42**(20):1353

[8] Kulakov VM, Ladygin EA. Effects of Ionizing Radiation on the Devices of Electronic Equipment. Sovetskoe Radio: Moscow, Russia; 1980. 244 p

[9] Srour JR, McGarrity JM. Radiation effects on microelectronics in space. Proceedings of the IEEE. 1988;**76**(11):1443-1469. DOI: 10.1109/5.90114

[10] Khattab K, Haddad K, Haj-Hassan H. Design of a permanent cd-shielded epithermal neutron irradiation site in the Syrian miniature neutron source reactor. Journal of Radioanalytical and Nuclear Chemistry. 2008;**277**(2):311-316. DOI: 10.1007/s10967-007-7081-7

[11] Anashin VS, Ishutin IO, Ulimov VN, Emeliyanov VV. Methods to control the hardness of specialized VLSI to space natural ionizing radiation. In: Stempkovsky A, editor. Problems of Perspective Micro- and Nanoelectronic Systems Development; October 4-8, 2010; Moscow Region. Moscow: IPPM RAS; 2010. pp. 233-236

[12] Fukuda M. Reliability testing of semiconductor optical devices. In: Ueda O, Pearton SJ, editors. Materials and Reliability Handbook for Semiconductor Optical and Electron Devices. 1st ed. New York: Springer; 2013. pp. 3-17. DOI: 10.1007/978-1-4614-4337-7_1

[13] Pecht MG, Chang MH. Failure mechanisms and reliability issues in LEDs. In: van Driel W, Fan X, editors. Solid State Lighting Reliability, Solid State Lighting Technology and Application Series. Vol. 1. New York, NY: Springer; 2013. pp. 43-110. DOI: 10.1007/978-1-4614-3067-4_3

[14] Dalapati P, Manik N, Basu A. Influence of temperature on the performance of high power AlGaInP based red light emitting diode. Optical & Quantum Electronics. 2015; **47**(5):1227-1238

[15] Alferov ZI. Nobel lecture: The double heterostructure concept and its applications in physics, electronics, and technology. Reviews of Modern Physics. 2001;**73**(3):767-782. DOI: 10.1103/RevModPhys.73.767

[16] Schubert EF, Cho J, Kim JK. Light-emitting diodes. In: Seidel A, editor. Kirk-Othmer Encyclopedia of Chemical Technology. John Wiley & Sons, Inc; 2015. pp. 1-20. DOI: 10.1002/0471238961.1209070811091908.a01.pub3

[17] Barnes CE. Effects of Co60 gamma irradiation on epitaxial GaAs laser diodes. Physical Review B. 1970;**1**(12):4735-4747. DOI: 10.1103/PhysRevB.1.4735

[18] Gromov DV, Maltcev PP, Nikiforov AY, Polevich SA, Startcev SA. Radiation hard GaAs microwave integrated circuits design. In: Radiation and Its Effects on Components and Systems, 1997. RADECS 97. Fourth European Conference on; September 15-19, 1997; Cannes, France. IEEE; 1997. pp. 147-149. DOI: 10.1109/RADECS.1997.698873

[19] Joint Stock Company "Research Institute of Devices". Research Simulator. The pulse solid-core dual-zone reactor on fast neutrons "BARS-4." [Internet]. Available from: http://www.niipriborov.ru/model_ustanov.html [Accessed: 06-07-2017]

[20] Figurov VS, Baykov VV, Shelkovnikov VV, Vasilev AV, Sushko MV. Estimation of effective spectral coefficient of neutron radiation in distant points of accelerator BARS-4 radiation

field. Questions of Atomic Science and Technics. Series: Physics of Radiation Effects on Radio-Electronic Equipment. 2011;**4**:39-51

[21] Gradoboev AV, Orlova KN, Asanov IA, Simonova AV. The fast neutron irradiation influence on the AlGaAs IR-LEDs reliability. Microelectronics Reliability. 2016;**65**:55-59. DOI: 10.1016/j.microrel.2016.07.143

[22] Gradoboev AV, Simonova AV. Influence of preliminary irradiation by gamma-quanta on development of catastrophic failures during operation of IR-LEDs. Journal of Physics: Conference Series. 2017;**830**:012132. DOI: 10.1088/1742-6596/830/1/012132

[23] Gradoboev AV, Sednev VV. Research on the radiation exposure "memory effects" in AlGaAs heterostructures. IOP Conference Series: Materials Science and Engineering. 2015; **81**(1):012007. DOI: 10.1088/1757-899X/81/1/012007

[24] Gradoboev AV, Sednev VV. The influence of power mode on IR-LED resistance to the irradiation with fast neutrons. Izvestiya Vysshikh Uchebnykh Zavedeniy. Fizika. 2014; **57**(10-3):20-23

[25] Lampert MA, Mark P. Current Injection in Solids. New York, NY: Academic Press; 1970. 363 p

[26] Gradoboev AV, Simonova AV, Orlova KN. Influence of irradiation by 60Co gamma-quanta on reliability of IR-LEDs based upon AlGaAs heterostructures. Physica Status Solidi (c). 2016;**13**(10-12):895-902. DOI: 10.1002/pssc.201600035

[27] Gradoboev AV, Orlova KN, Asanov IA. Radiation model of the LEDs based on heterostructures AlGaInP. Irradiation by gamma-quanta 60Co. In: Bondarenko GG, editor. XXII International Conference "Radiation Physics of Solid State"; July 9-14, 2012; Sevastopol. Moscow: NIIPMT-MIEM HSE; 2012. pp. 510-516

[28] Gradoboev AV, Orlova KN, Asanov IA. Irradiation of LEDs based on AlGaInP heterostructures with multiple quantum wells by 60Co gamma-rays. Perspektivnye Materialy. 2013;**7**:49-55

[29] Gradoboev AV, Orlova KN, Simonova AV. E-MRS Spring Meeting 2017, Strasbourg, France. Symposium O.7.14· Parameters changes of LEDs based on GaP under irradiation by gamma-quanta [Internet]. February 27, 2017. Available from: http://www.european-mrs.com/wide-bandgap-semiconductors-leds-solar-and-related-energy-technologies-emrs#collapse43 [Accessed: 11-07-2017]

[30] Barnes CE, Soda KJ. Application of damage constants in gamma irradiated anphoterically Si doped GaAs LEDs. IEEE Transactions on Nuclear Science. 1976;**23**(6):1664-1670. DOI: 10.1109/TNS.1976.4328559

[31] Barry AL, Houdayer AJ, Hinrichsen PF, Letourneau WG, Vincent J. The energy dependence of lifetime damage constants in GaAs LEDs for 1-500 MeV protons. IEEE Transactions on Nuclear Science. 1995;**42**(6):2104-2107. DOI: 10.1109/23.489259

[32] Gradoboev AV, Rubanov PV, Ashcheulov AV. Russian Federation Patent for an Invention no.2303314. IPC: H01 L21/18. Method for manufacturing semiconductor devices. Federal Institute of Industrial Property. Declared April 17, 2006. Published July 20, 2007; Bulletin no. 20. 6 p

[33] Gradoboev AV. Study of the "memory effect" in 3-cm band Gunn diodes under irradiation by fast neutrons. In: Microwave & Telecommunication Technology (CriMiCo), 2014 24th International Crimean Conference; September 7-13, 2014; Sevastopol, Ukraine. IEEE; 2014. pp. 870-871

Chapter 2

Applicability of Quantum Dots in Biomedical Science

Slavica Brkić

Additional information is available at the end of the chapter

http://dx.doi.org/10.5772/intechopen.71428

Abstract

Quantum dots (QDs) are novel class of inorganic fluorophore with superior photophysical properties. Superior optical properties are a promising alternative to organic dyes for fluorescence biomedical applications. These nanoparticles have size-tunable emission, strong light absorbance, and very high levels of brightness and photostability. Highly luminescent QDs are prepared by coating the core with another material, resulting in core-shell quantum dots that are more stable in various chemical environments. These core-shell QDs are hydrophobic and only organic soluble as prepared. Hydrophobic QDs are insoluble in aqueous solution and cannot be directly employed in biomedical applications. They are necessarily made water soluble by surface modifying them with various bifunctional surface ligands or caps to promote aqueous solubility and enhancing biocompatibility. To make them useful for biomedical applications, QDs need to be conjugated to biological molecules without disturbing the biological function of these molecules. Most of the current studies were designed to ask questions concerning the physicochemical properties of novel QD products, not QD toxicity *per se*. The potential toxicity of the QDs is a cause for concern because they are made of heavy metals. The limitation of heavy metal–containing QDs stimulates extensive research interests in exploring alternative strategies for the design of fluorescent nanocrystals with high biocompatibility.

Keywords: quantum dot, band gap, core/shell/ligand structure, hydrophobic QD, biocompatibility, hydrophilic QD, functionalization, toxicity

1. Introduction

Quantum dots (QDs) are semiconductor nanoparticles that are restricted in three dimensions, typically with a diameter of 2–8 nm [1]. These particles are defined as particles with physical dimensions smaller than the exciton Bohr radius [2]. QDs are a bridge between bulk materials and atomic or molecular structures and characterized by composition-dependent band gap energy. These particles are light-emitting nanocrystals with novel optical and electrical properties. Properties of QDs have attracted great interest in biology and medicine in the recent years [3]. Compared with organic dyes and fluorescent proteins, semiconductor QDs offer several unique advantages. However, the major concerns about potential toxicity of QDs have cast doubts on their practical use in biology and medicine [3].

2. The band gap energy

The small size of QDs lead to what is known as "quantum confinement" [4]. The quantum confinement effects occur when size of nanoparticle smaller than exciton Bohr radius. An exciton Bohr radius is the distance in an electron-hole pair in a bulk semiconductor. QDs are defined as particles with physical dimensions smaller than the exciton Bohr radius. The quantum confinement means that the energy levels that the electrons inhabit become discrete, with a finite separation between them. However, there are some energy levels that the electrons cannot occupy, which are collectively known as band gap [5]. Most electrons occupy energy levels below this band gap in the area known as the valence band, indeed most energy levels in the valence band are occupied. If, however, an external stimulus is applied, an electron may move from the valence band to the conduction band, i.e., those energy levels above the band gap. When the QD is hit by incident light, it absorbs a photon with a higher energy than that of the band gap of the composing semiconductor. When the electron returns to a lower energy level, a narrow, symmetric energy band emission occurs [6]. The wavelength of photon emissions depends not only on the material from which the dot is made but also its size; the smaller the size of the QDs, the larger the band gap energy and QDs emit blue light, larger QDs having smaller band gaps emit the larger wavelength. Recombination occurs when an electron from a higher energy level relaxes to a lower energy level and recombines with an electron hole. This process is accompanied by the emission of radiation, which can be measured to give the band gap size of a semiconductor. The energy of the emitted photon in a recombination process of a QD can be modeled as the sum of the band gap energy, the confinement energies of the excited electron and the electron hole, and the bound energy of the exciton [1]:

$$E = E_{bandgap} + E_{confinement} + E_{exciton} \tag{1}$$

The confinement energy can be modeled as a simple particle in an one-dimensional box problem and the energy levels of the exciton can be represented as the solutions to the equation at the ground level ($n = 1$) with the mass replaced by the reduced mass. The magnitude of this confinement energy:

$$E_{confinement} = \frac{\hbar^2 \pi^2}{2 d^2}\left(\frac{1}{m_e} + \frac{1}{m_h}\right) = \frac{\hbar^2 \pi^2}{2\mu d^2}, \tag{2}$$

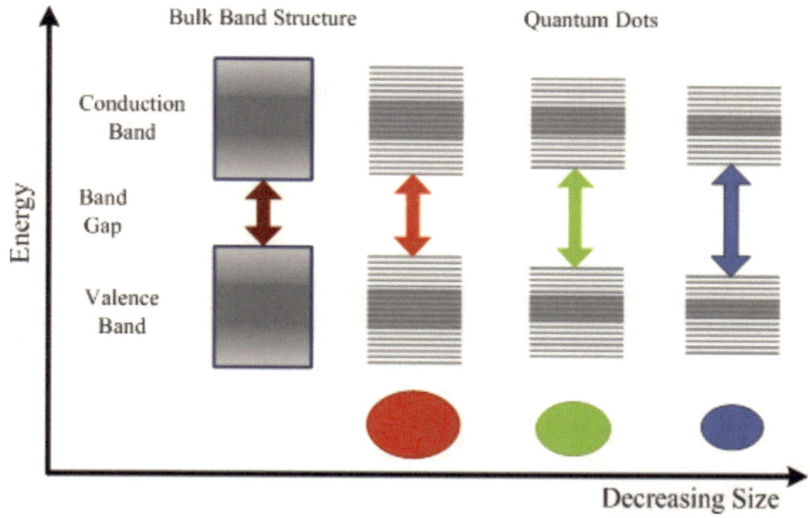

Figure 1. Size-dependent band gap energy [2].

where m_e is the effective mass of the electron, m_h is the effective mass of the hole, μ is the reduced mass of the exciton system, and d is diameter of the confinement [7]. The bound exciton energy can be modeled by using the Coulomb interaction between the electron and the positively charged hole. The negative energy is proportional to Rydberg's energy (R_y = 13.6 eV) and inversely proportional to the square of the size-dependent dielectric constant, ε_r. Energy of exciton:

$$E_{exciton} = -\frac{1}{\varepsilon_r^2}\frac{\mu}{m_e}R_y. \tag{3}$$

Using these models and spectroscopic measurements of the emitted photon energy E, it is possible to measure the band gap of QDs. Decreasing the size of a QD results in a higher degree of confinement, which produces an exciton of higher energy, thereby increasing the band gap energy as can be seen in **Figure 1**.

3. Biomedical application of quantum dots

Nanotechnology has been heralded as a new field that has the potential to revolutionize medicine, as well as many other seemingly unrelated subjects, such as electronics, textiles, and energy productions. In 1998, the potential of these nanoparticles for applications involving biological labeling was first reported. The first successful application of QDs to medical diagnostics has been demonstrated by immunofluorescent labeling of fixed cells and tissues and immunostaining of membrane proteins on living cells [8]. The fact that several QDs can be excited by the same wavelength of light opens up several multiplexing potentials, including high-throughput screening of biological samples [9]. Size-tunable absorption and emission property of QDs is an extremely valuable property for biological imaging as they can be tuned all the way from the UV to the near-infrared of the spectrum. For biological and medical applications, it is of importance to study the photophysical properties of QDs.

3.1. Photophysical properties of quantum dots

In the biomedical sciences, fluorescence is used as a powerful tool for labeling, imaging, tracking, detection, and therapy. The fluorescent labeling of biological molecules using organic fluorophore and QDs. QDs are new class of fluorophores that offer several advantages over traditional fluorophores. These particles have broad absorption and narrow emission spectra, high quantum yield, long life-time, high brightness, and stable against photobleaching. A great advantage of QDs compared to the classical fluorophore which emits light in the infrared and near-infrared regions, as the absorption of tissues is minimal in this region. The most popular types of QDs include CdSe, CdTe, and ZnSe. The most commonly studied and used QD is cadmium selenide. The broad absorption spectra of the QDs allows single wavelength excitation of emission from different-sized QDs. Multicolor optical coding for biological assays has been achieved by using different sizes of quantum dots with precisely controlled ratios [10]. In order to apply QDs in biomedical imaging, recent studies have focused on developing near-infrared luminescent QDs which exhibit an emission wavelength ranging from 700 to 900 nm [3]. The use of the near-infrared (NIR) photons is promising for biomedical imaging in living tissue due to longer attenuation distances and a better tissue staining without interfering with autofluorescence, since most of the tissue chromophores weakly absorb light in the infrared range of wavelengths. Hemoglobin and water have lower absorption coefficient and scattering effects in the NIR region (650–900 nm). The QDs emission can be set to a NIR area by adjusting QD size or by incorporating rare-earth activators. The unique robust optical properties of QDs and their surface properties that allow biocompatibility and heteroconjugation make QDs highly promising fluorescent labels for biological applications with significant superiority over classic organic fluorophores [10].

3.1.1. Size-dependent optical properties

Quantum dots exhibit size-dependent discrete energy levels. The energy gap increases with decrease in the size of the nanocrystal, thus yielding a size-dependent rainbow of colors. The wavelength of the emission photon depends not on the material from which the dot is made but on its size. The ability to control the size of QD enables the manufacturer to determine the wavelength of emission, which in turn determines the color of light the human eye perceives. The smaller the dot is closer to the blue end of the spectrum, and the larger the dot is closer to the red (**Figure 2**) [9]. Light wavelengths from ultraviolet to infrared region (400–4000 nm) can be achieved with variation of the size and composition of nanoparticles [11]. The optical properties of QDs will change with proximity of quantum dots to each other.

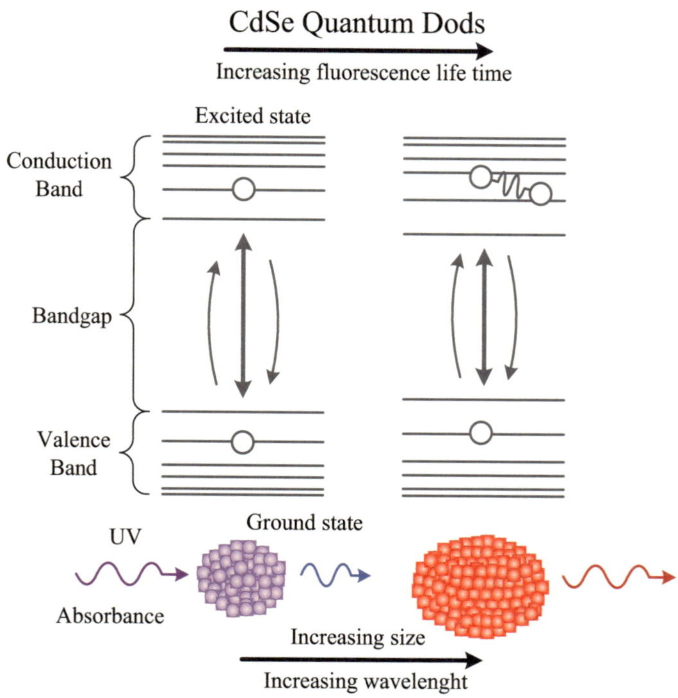

Figure 2. Band gap energy and size dependent wavelength of CdSe quantum dot [12].

Stokes shift

One of the most common features of QDs is the photoluminescence redshift relative to absorption, also called Stokes shift. It was named after Irish physicist George G. Stokes. Stokes shift is the difference between QD's peak excitation and the peak emission wavelengths (**Figure 3**) [5]. The energy associated with emission is typically lower than the excitation light. The redshift of emission peaks, with respect to absorption spectra, is size dependent. Stokes shift is commonly observed in semiconductor QDs, and is one of the most important quantities that determine the optical properties of QDs. The large separation between the excitation and emission spectra of the QDs improves the detection sensitivity, as the entire emission spectra of QDs can be detected. The Stokes shift of semiconductor QDs can be as large as 300–400 nm, depending on the wavelength of the excitation light. As the radius of quantum dot increases, the redshift decreases and disappears beyond a certain radius. Each QD has a unique emission maximum; changing the dot size leads to the emission maximum shift. In addition, QDs have very broad absorption spectra, and can be excited over the entire visual wavelength range as well as far into ultraviolet [13]. Because of their exceptionally large Stokes shifts of up to 400 nm, QDs can be used for the multicolor detection with a single wavelength excitation source (**Figure 4**) [14].

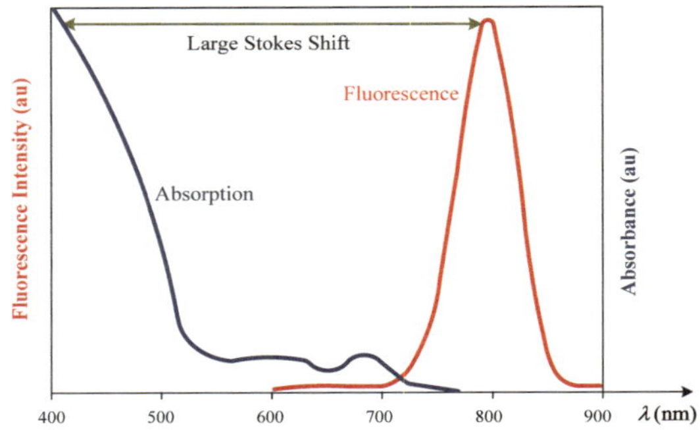

Figure 3. Broad absorption and narrow emission spectrum [15].

Fluorescence intensity and lifetime

The fluorescence of QDs is generated when the excited electron emits photon and returns to the ground-state. The lifetime is a delay between the moment of absorbtion a photon from the light sorce and moment of emitted light (**Figure 5**). The lifetime of larger dots have more closely spaced energy levels in which the electron-hole pair can be trapped. Therefore, electron-hole pairs in larger dots live longer [10]. Long lifetime provides difference of QD fluorescence signal from background fluorescence. For example, see [17]. The life time of QDs and fluorophores is shown in **Figure 6**.

In order to extend the lifetime of the QDs, rare earths can be incorporated into the QDs which create local quantum states.

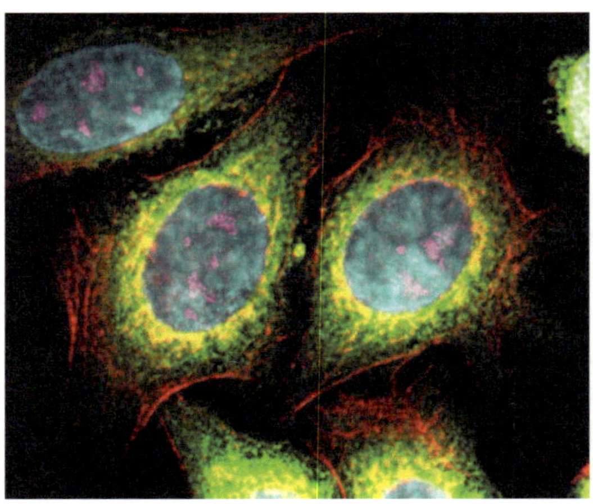

Figure 4. Color of light depends on size quantum dot.

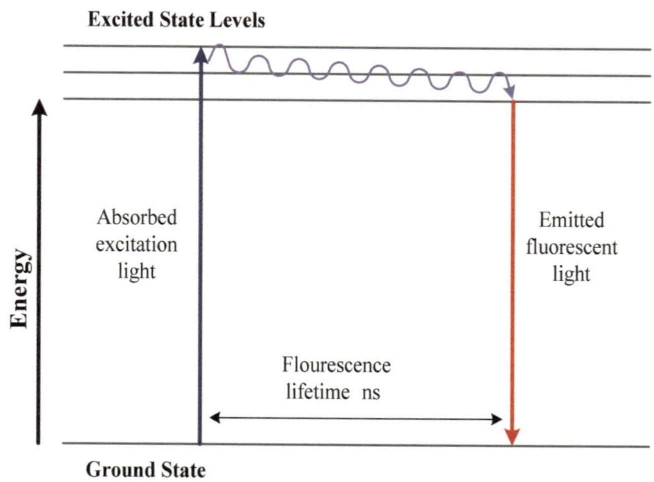

Figure 5. Fluorescence lifetime [2].

Brightness and photostability

Minimizing overlap between total excitation and emission bands enhances the clarity and brightness of the fluorescing QD by avoiding re-absorption of emitted light into nearby quantum dots—characteristics that display manufacturers and end-users find highly desirable. Broad absorption spectra make it possible to excite all QDs simultaneously with a single light source and minimize sample autofluorescence by choosing an appropriate excitation wavelength. QDs have very large molar excitation coefficient in the order of 0.5–5 × 10^6 M^{-1}, about 10–50 times larger than that of organic dyes. QDs are able to absorb 10–50 times more photons than organic

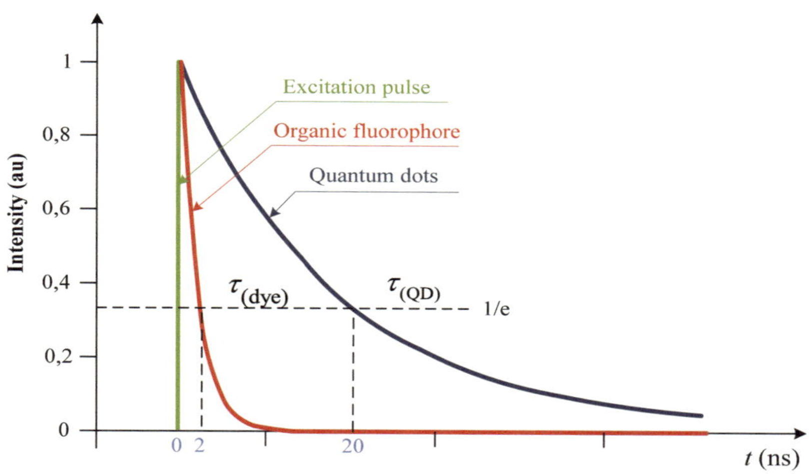

Figure 6. Fluorescence lifetime quantum dots and organic fluorophore [19].

dyes at the same excitation photon flux, leading to a significant improvement in the probe brightness. Owing to their inorganic nature, QDs have minimal interaction with the surrounding environment which contributes to their photostability. Inorganic QDs are more photostable under ultraviolet excitation than organic molecules, and their fluorescence is more saturated [18].

Quantum yield

Quantum yield is a measure of the "brightness" of a fluorophore and is defined as the ratio of the number of photons emitted to the number of photons absorbed. Some organic dyes have quantum yields approaching 100%, but conjugates (from biological affinity molecules) made from these generally have a significantly lower quantum yield. QDs are relatively efficient with regards to conversion of the excitation light into emission, where the quantum yield is generally over 50%. QDs retain their high quantum yield even after conjugation to biological affinity molecules [14]. The fluorescence quantum yield gives the efficiency of the fluorescence process.

Photobleaching

Photobleaching (fading) is the photochemical alteration of a dye or a fluorophore molecule, such that it is permanently unable to fluoresce. This is caused by cleaving of covalent bonds or non-specific reactions between the fluorophore and surrounding molecules. QDs are several thousand times more stable against photobleaching than organic dye molecules, and are thus ideal probes for fluorescent spectroscopy and bioimaging applications [20].

4. Core/shell structure quantum dot

Quantum dots are used as bare core or as core/shell structures. Although these "core" QDs determine the optical properties of the conjugate, they are by themselves unsuitable for biological probes owing to their poor stability and quantum yield [21]. In fact, the quantum yield of QD cores has been reported to be very sensitive to the presence of particular ions in solution [22]. A bare nanocrystal core is highly reactive and toxic, resulting in a very unstable structure that is prone to photochemical degradation. The core is highly reactive due to their large surface area/volume ratio, resulting in a very unstable structure which is particularly prone to photochemical degradation. Capping the core with a semiconductor material of a higher band gap not only increases the stability and quantum yield, but also passivates the toxicity of the core by shielding reactive ions from being exposed to photo-oxidative environments, e.g., exposure to UV and air (**Figure 7**). Traditionally, the typical QDs consist of a II–IV, IV–VI, or III–V semiconductor core (CdTe, CdSe, Pb, Se, GaAs, GaN, InP, and InAs), which is surrounded by a covering of wide band gap semiconductor shell as zinc sulfide ZnS or cadmium sulfide CdS is generally used (approx. 1.5 nm thick) [23]. Capping core nanocrystals with ZnS has shown to increase stability and performance, producing QDs with improved photophysical and chemical properties at room temperature. However, ZnS capping alone is not sufficient to stabilize the core, particularly in biological solutions [24]. Quantum dots do not exhibit aqueous solubility as they are generally synthesized in organic solution. The

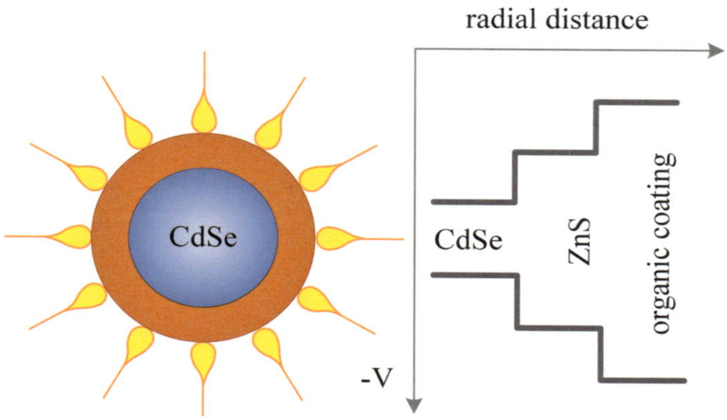

Figure 7. Band gap energy core/shell/ligand structure [12].

outmost surface of these nanoparticles is passivated with an organic ligand.[1] QDs are made water soluble by surface modifying them with various caps to ensure aqueous solubility and enhancing biocompatibility. The three important features of QDs used in the display industry, called core-shell QDs, are the core, shell, and ligand (**Figure 8**).

When these three features of QDs are tuned to how we need them, we develop exciting new applications. In order to utilize in a biological environment, they need to be made hydrophilic. Thus, they are necessarily made water soluble by surface modifying them with various bifunctional surface ligands or caps to promote aqueous solubility and enhancing biocompatibility [25]. QDs must be conjugated with molecules which have the capabilities of recognizing the target. These surface modifications can also help prevent aggregation, reduce the nonspecific binding, and are critical in achieving specific target imaging in biological studies [26].

4.1. Surface coatings and water-solubility

The key to develop QDs as a tool in biological systems is to achieve water solubility, biocompatibility, and photostability [27]. QDs, single or core/shell structures, do not exhibit aqueous solubility. After synthesis, hydrophobic QD must be covered with an organic layer or incorporated within the organic shell to make them water-soluble and biocompatible [28]. Some of the techniques used to achieve solubilization include ligand exchange, surface silanization, and phase transfer method [29].

4.1.1. The ligand exchange method

Native cap exchange is a strategy in which the native hydrophobic layer (TOPO/TOP)[2] cap is replaced with a bifunctional moiety that can bind QDs from one side while exposing hydrophilic

[1]As organic ligands, molecules and ions, containing O, S, N, and P, are bonded so that their electronic pair can produce a covalent bond with the central atom.
[2]TOPO/TOP: Typically trioctylphosphine oxide/trioctylphosphine.

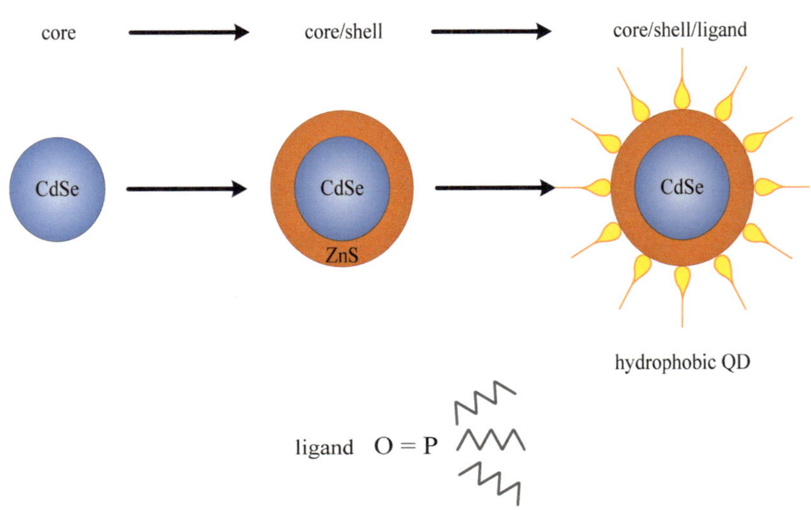

Figure 8. Core/shell/ligand structure of hydrophobic quantum dot [12].

groups on the surface to achieve optimal dispersion [30]. The ligand exchange method includes the exchange of the hydrophobic surfactant molecules with bifunctional molecules, which are hydrophilic on one side and hydrophobic on the other, to bind to the ZnS shell on the QD [23] (**Figure 9**). These biocompatible polymers usually have functional anchor groups, such as thiol, amine, and carboxyl, which can passivate QDs more strongly than the original ligand [31]. Most often thiols (–SH) as functional groups are used to bind to the ZnS and carboxyl (–COOH) groups are used as hydrophilic ends. The resulting QDs are soluble in both aqueous and polar solvents [29]. Biomolecules, such as proteins, peptides, and DNA, were also conjugated to the free carboxyl groups by crosslinking to the reactive amine. This process did not affect the optical characters of the QDs compared with the original QDs.

4.1.2. Surface silanization

Surface silanization involves the growth of a silica shell around the nanocrystal. As silica shells are highly cross-linked, they are very stable [32]. Silica coating enhances the mechanical stability of colloidal QDs and protects them against oxidation and agglomeration. The adventure of silica encapsulation is QDs chemical stability over a much broader pH range compared to carboxy-terminated ligands [29]. The chemistry of glass surfaces can be readily extended to silanized QDs, providing more flexibility for bioconjugation (**Figure 9**).

4.1.3. Polymer coating (phase transfer)

Polymer coating (phase transfer) method uses amphiphilic polymers to coat the surface [33, 34]. The hydrophobic alkyl chains of the polymer interdigitate with the alkyl groups on the QDs surface, while the hydrophilic groups orientate outwards to attain water solubility. However, coating with a polymer may increase the overall diameter of the QDs, and this

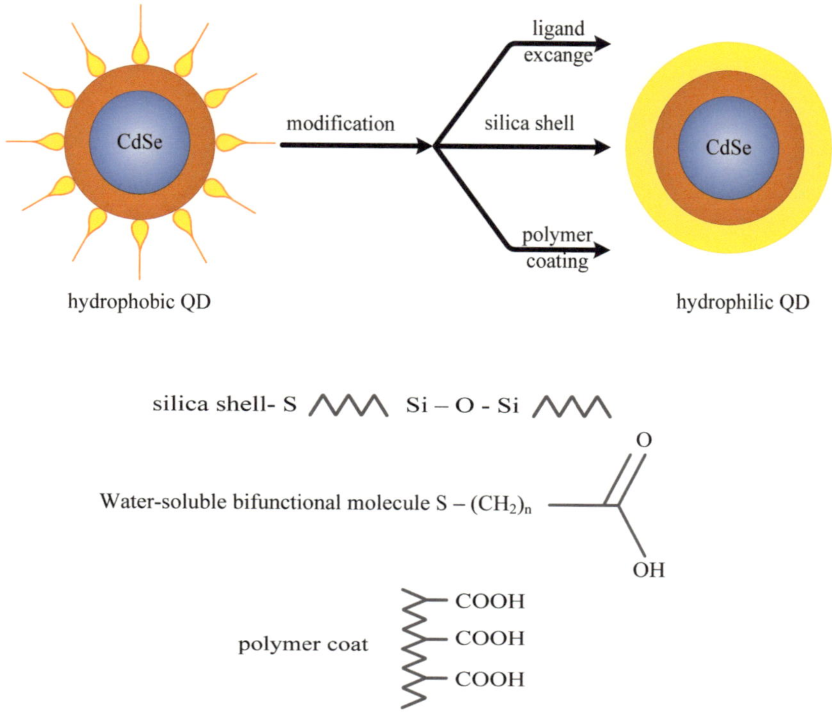

Figure 9. Modification from hydrophobic to hydrophilic quantum dots [12].

may reduce emission and limits their use biological applications [35]. The aqueous coating can then be tagged with various biomolecules of interest. Biomolecules, such as proteins, peptides, and antibodies, were conjugated to the free functional groups by crosslinking to the reactive amine. This process did not affect the optical characters of the QDs compared with the original hydrophobic surfactant layer.

5. Functionalization of quantum dots

Once solubilization has been achieved, QDs can be functionalized by conjugation to a number of biological molecules. QDs are adapted to the desired biological application by conjugation to biological molecules without disturbing the function of these molecules. QDs must be conjugated with molecules which have the capabilities of recognizing the target [36]. Biomolecules include antibodies, peptides, avidin, biotin, oligonucleotides, albumin, or by coating with streptavidin (**Figure 10**) [6]. Owing to large surface area-to-volume ratio, several biomolecules of varying types can be attached to a single QD. Each of these biomolecules provides a desired function which affords multi-functionality [37]. Various surface modification techniques were developed to ensure the specific bioconjugation: covalent linkage, electrostatic attraction, adsorption, and mercapto (–SH) exchange. The choice depends on the features of the biomolecule of interest

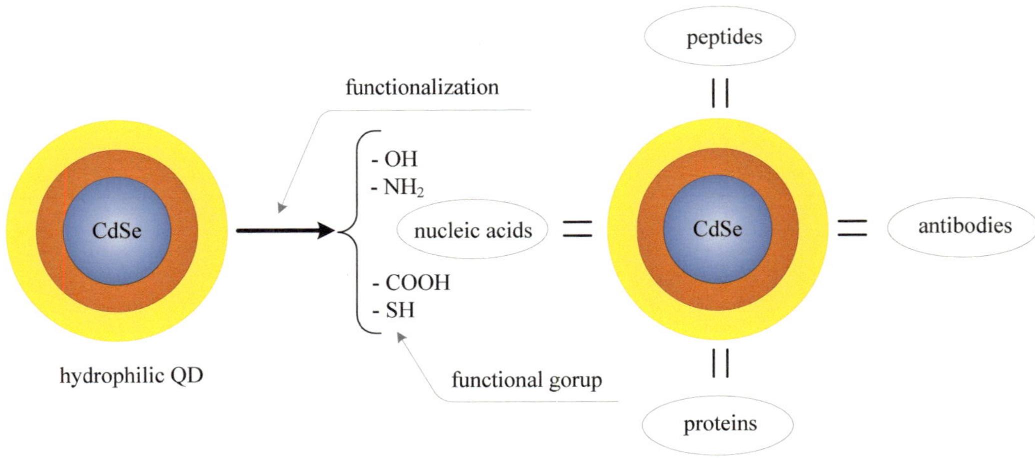

Figure 10. Functionalization of hydrophilic quantum dots [16].

[38]. Surface modifications can also help prevent aggregation, reduce the nonspecific binding, and are critical to achieving specific target imaging in biomedical studies [3]. QDs retain excellent stability of optical properties upon conjugation to biomolecules, and can be simultaneously excited by a single light source.

5.1. Covalent linkage

The first bioconjugation method offers a most stable covalent linkage of QDs to biomolecules and is the most commonly used approach for making biofunctionalized QDs for *in vivo* applications. Most water solubilization methods result in QDs covered with carboxylic acid, amino or thiol groups. Under these situations, it is easy to link QDs to biological molecules which also have these functional groups [39]. Using these methods, there have been numerous reports of conjugating QDs with various biological molecules, including proteins, antibodies, peptides, oligonucleotides, and albumin [5].

5.2. Electrostatic interaction

In the second procedure, an electrostatic interaction between QDs and charged adapter molecules or proteins with incorporated charged domains was employed [40, 41]. In this case, the protein of interest can be fused to a positively charged domain that will in turn bind electrostatically to the negatively charged surface of the QDs [6]. In the electrostatic interaction approach, the binding energies are highly dependent on both the chemical environment and the ambient temperature and QDs size can strongly affect the interaction efficiency. Electrostatic interactions generally are not sufficiently specific, however, given the complexity of biological milieu [5]. Therefore, conjugates made by this way are not suitable for *in vivo* or *ex vivo* cell labeling due to the possible interference with positively charged proteins [42].

5.3. Non-specific adsorption

Simple small molecules such as oligonucleotides and various serum albumins were found to readily adsorb to the surface of water-soluble QDs [6]. This adsorption is nonspecific and depends on ionic strength, pH, temperature, and surface charge of the molecule [5].

5.4. Mercapto exchange

Thiol-containing biomolecules can be conjugated to QDs via mercapto exchange. Thiols (SH) bond of mercapto group bind to the surface of the most often used semiconductor materials (CdSe, CdS, CdTe, and ZnS), and therefore QDs can be conjugated to biological molecules bearing mercapto (SH) groups in this way [43]. ZnS-thiol bond is not very strong. The conjugated biomolecules may break off and the QDs may precipitate out of solution [6].

Surface of QDs are decorated with different biomolecules: proteins, peptides, nucleic acids, ligands, or other biomolecules to achieve bioconjugation that mediate specific interactions with living tissue.

6. Toxicity of quantum dots

Most of the current studies were designed to ask questions concerning the physicochemical properties of novel QD products such as fluorescence, detectability, stability, and cell labeling efficacy, and not QD toxicity *per se*. Several studies reveal that the toxicity of QDs depends on many factors which can be summarized as inherent physicochemical properties and environmental conditions. Understanding the potential toxicity of QDs requires a fundamental grasp of QD physicochemical properties. Each QD type will need to be characterized individually to its potential toxicity. Each individual type of QD possesses its own unique physicochemical properties, which in turn determines its potential toxicity or lack thereof. QD sizes, charges, concentrations, outer coating materials and functional groups, oxidation, and mechanical stability all have been implicated as contributing factors to toxicity [44]. Most QDs consist of heavy metal which may be potentially toxic. Possibly the most important aspect of QD toxicity is their stability, both *in vivo* and during synthesis and storage. QDs with an outer ZnS shell, referred to as core-shell QDs, are generally more chemically stable. Degradation of the QD coating may also result in reaction of the QD in undesirable/unanticipated ways *in vivo*. Further, some QD coating materials have themselves been found to be cytotoxic. Cytotoxicity of QDs has been observed in a large number of *in vitro* studies [45], affecting cell growth and viability [22]. Several studies suggest QD cytotoxicity to be due to photolysis or oxidation. Under oxidative and photolytic conditions, QD core-shell coatings have been found to be labile, degrading, and thus exposing potentially toxic "capping" material or intact core metalloid complexes or resulting in dissolution of the core complex to QD core metal components, e.g., Cd and Se [46]. The toxicity is associated with the cadmium, lead, or arsenic containing QDs. For *in vivo* studies, the main concern is the robustness of the surface coating.

An unstable surface coating could expose the core of the QD to UV damage or air oxidation. The toxic heavy metal ions can be easily leaked out into biological systems, if the surfaces are not properly covered by the shells or protected by ligands [44]. The stability of QDs and their resistance to metabolic degradation in live cells would allow long-term imaging studies, and several studies have indicated lack of cytotoxicity for period up to 4 months [47]. Although such QDs should not be acutely toxic as long as their polymer coating is stable enough to restrain the release of cadmium, both short- and long-term safety of QDs will need to be established in toxicological studies in clinically relevant animal models. Studies in cell lines have shown that QDs do not affect cell growth under normal media conditions and short-term administration of QDs into animals seems not to affect the metabolism and behavior of the animals [48]. Xiao and coworkers showed that the cytotoxicity of CdSe quantum was proportional to QDs concentration in the tested range. There was a negligible cytotoxicity at concentration of 20 nM, which is twice of the applied concentration of fluorescent cell imaging test. On the other side, no size-dependent cytotoxicity was observed. For time-dependence, cytotoxicity increased along with prolongation of incubation period, and tended to be stable after 12 h of incubation. Therefore, cytotoxicity can be neglected in a typical fluorescent cell imaging procedure (10 nM/L, 12 h), and the labeling effect to microtubule can be guaranteed satisfactorily [49]. There are also concerns about toxicity related to the molecules used for surface functionalization of the QDs [6]. However, toxicity concerns are still valid for longer periods and *in vivo* applications. Given the highly intrinsic toxicity of cadmium, the biological applications of Cd-QDs have been limited. The limitation of heavy metal containing QDs stimulates extensive research interests in exploring alternative strategies for the design of fluorescent nanocrystals with high biocompatibility. In this case, the practical strategy is to develop highly fluorescent nanoparticles based on nontoxic elements [3].

6.1. Nontoxic quantum dots

The last few years of research are focused on QDs that do not contain Cd. QDs made up of III–V semiconductors such as InP. Groups III–V QDs may provide a more stable alternative to groups II–VI (CdSe, Cd, Te, etc.) [50]. QDs due to the presence of a covalent, rather than an ionic bond, and have been reported to have lower cytotoxicity. However these QDs are difficult to prepare on a competitive time scale, and tend to have much lower quantum efficiencies, meaning uptake has been slow [51]. The most highlighted candidates for Cd-free QDs are the InP-based QDs. InP QDs a very attractive candidate for replacing cadmium-based QDs for biological studies. However, till date, there are very little studies reporting the use of InP QDs in bioimaging applications, as they are difficult to prepare because of the sensitivity of precursors and surfactants toward the reaction environment in obtaining good quality InP QDs [52]. Therefore, progresses in the synthesis of InP-based QDs occurred much slowly compared to that of CdSe QDs [53]. InP/ZnS QDs combining the brightness, photostability, and biocompatibility characteristics will serve as a new generation of targeted optical probes for several biomedical applications, including early detection of cancer, replacing the cadmium-based [54]. Silicon nanoparticle is another type of QDs which possess unique properties when their sizes are reduces to below 10 nm. Similar to their predecessors, silicon QDs also have many advantages over traditional fluorescent organic dye. Moreover, owing to silicon's nontoxic and environment-friendly nature, Si QDs are used as fluorescent probes for bioimaging. Si QDs are produced in nonpolar solvents with hydrophobic ligands on their surface in other to protect the Si core. Therefore, it is a common problem that silicon QDs show poor solubility and unstable photoluminescence in aqueous solutions. Thus, fabricating good water-disperse Si QDs with stable optical characters is vital to their applications in bioimaging studies. In the demand of using biocompatible and nontoxic QDs as nanoprobes, rare-earth (RE) elements are used to fabricate a new type of QDs, such as Gd-doped ZnO QDs [44].

7. Adventures of quantum dots compared to conventional fluorophores

The ability to study molecular and cellular events by using fluorescent probes has broadly impacted many areas in biomedical research. QDs offer several adventures over traditional fluorophores. Traditional fluorophores have certain properties that limit their biomedical applications: narrow absorption and broad emission spectra, low luminescence and quantum yield, rapid photobleaching, and low brightness signal. A limitation of traditional small-molecule fluorescent dyes is in the labeling of other small molecules, drugs, transporters, and small-molecule probes to cell-surface receptors. Conjugates of dyes to these small molecules often lack sensitivity or specificity in the detection of the desired targets. Conjugates of small molecules to QDs produce conjugates with much greater light output per binding event, owing to the increased absorbance and emission of the QD [14]. The fact that several QDs can be excited by the same wavelength of light opens up several multiplexing potentials, including high-throughput screening of biological samples [9]. Size-tunable absorption and emission property of QDs is an extremely valuable property for biological imaging as they can be tuned all the way from the UV to the near-infrared of the spectrum. QDs exhibit better photophysical properties than conventional fluorophores under appropriate conditions. A limitation of traditional small-molecule fluorescent dyes is in the labeling of other small molecules, drugs, transporters, and small-molecule probes to cell-surface receptors. Conjugates of dyes to these small molecules often lack sensitivity or specificity in the detection of the desired targets. Conjugates of small molecules to QDs produce conjugates with much greater light output binding event, owing to the increased absorbance and emission of the QD. Furthermore, there is the possibility of improved avidity compared of a dye conjugate, owing to the combined effect of many molecules of the binding ligand on the surface of the quantum dot [14]. Biocompatible QDs represent a powerful tool for the direct readout of information down to single molecule level [28]. For example, by conjugating an antibody to the QD, targeting to specific cells or tissues can be affected *in vivo* or in fluorescent antibody immunoassays. QDs have shown great potential to provide spatial, temporal, and structural information for biological systems: *in vivo* cell labeling, *in vitro* cell labeling, detection of tumor marker, *in situ* tissue diagnostic, noninvasive tumor imaging, tracking of local cancer grown and its distant dissemination, detection and therapy of various disease, identifying of various types of biomarkers, more effective and early diagnosis of cancer, imaging of live tissue, etc.

8. Conclusion

Quantum dots have emerged as a new promising class of fluorescent probes for biomolecular and cellular imaging. In comparison with organic dyes and fluorescent proteins, QDs have unique optical and electronic properties. QDs do not exhibit aqueous solubility as they are generally synthesized in organic solution and are surface-stabilized with hydrophobic organic ligands. To make them useful for biomedical applications, QDs need to be conjugated to biological molecules without disturbing the biological function of these molecules. Various surface modification techniques were developed to ensure the specific bioconjugation. This is usually achieved by decorating QDs with proteins, peptides, nucleic acids, streptavidin, or other biomolecules that mediate specific interactions with living systems. Biocompatible quantum dots represent a powerful tool for the direct readout of information down to single molecule level. The limiting factor of application QDs *in vivo* is their toxicity. Although QD technology is still not in much use due to their hydrophobicity, toxicity, and many issues need to be solved in order to apply them safely in biomedical applications.

Author details

Slavica Brkić

Address all correspondence to: slavica.brkic@sve-mo.ba

Faculty of Science and Education, University of Mostar, Mostar, Bosnia and Herzegovina

References

[1] Mahajan SV et al. Characteristics and properties of CdSe quantum dots. International Journal of Latest Research in Science and Technology. 2013;**2**(1):457-459. ISSN (Online): 2278-5299

[2] Wang HZ et al. Detection of tumor marker CA125 in ovarian carcinoma using quantum dots. Acta Biochimica et Biophysica Sinica. 2004;**36**(10):681-686. [PubMed]

[3] Jin S et al. Application of quantum dots in biological imaging. Journal of Nanomaterials. 2011;**2011**:834139

[4] West JL, Halas NJ. Engineered nanomaterials for biophotonics applications: Improving sensing, imaging, and therapeutics. Annual Review of Biomedical Engineering. 2003; **5**:285-292

[5] Xing Y, Jianghong R. Quantum dot bioconjugates for *in vitro* diagnostic & *in vivo* imaging. Cancer Biomarkers. 2008;**4**:307-319

[6] Azzazy HME et al. From diagnostic to therapy: Prospects of quantum dots. Clinical Biochemistry. 2007;**40**:917-927

[7] Dey S, Jain YS. On the wave mechanics of a particle in two different impenetrable spherical cavities. arXiv:1002.4308v1, 2010

[8] Sukhanova A et al. Highly stable fluorescent nanocrystals as a novel class of labels for immunohistochemical analysis of paraffin-embedded tissue sections. Laboratory Investigation. 2002;**82**:1259-1261

[9] Smith AM et al. Multicolor quantum dots for molecular diagnostics of cancer. Expert Review of Molecular Diagnostics. 2006;**6**:231-244

[10] Bera D et al. Quantum dots and their multimodal applications. Materials. 2010;**3**:2260-2345. DOI: 10.3390/ma3042260

[11] Pietryga J et al. Pushing the band gap envelope: Mid-infrared emitting colloidal PbSe quantum dots. Journal of the American Chemical Society. 2004;**126**(38):11752-11753

[12] Brkić, S., Biocompatibility of cadmium selenide quantum dots, European International Journal of Science and Technology. 2017;**6**(4):6-17

[13] Dubrovsky T. Semiconductor nanoparticles as reporters in multiplexed immunoassay and cell analysis. International Journal of Nanoscience. 2009

[14] Hotz ZC. In: Rosenthal SJ, Wright DW, editors. Application of Quantum Dots in Biology, Nano Biotechnology Protocols. Nashville, TN: The Department of Chemistry, Vanderbilt University; 2005

[15] Quantum Dots - BME 240 [Internet] 2007. Available from: http://bme240.eng.uci.edu/students/07s/yokabe/advantages.htm

[16] Qdot Imaging MAT 594-20085. [Internet] Available from: https://www.mat.ucsb.edu/~g.legrady/academic/courses/08s594/rch/pc/pc.html

[17] Dubertret B, Skourides P, Norris DJ, Noireaux V, Brivanlou AH, Libchaber A. *In vivo* imaging of quantum dots encapsulated in phospholipid micelles. Science. 2002;**298**:1759-1762

[18] Weissleder R. A clearer vision for *in vivo* imaging. Nature Biotechnology. 2001;**19**:316-317

[19] Jamienson T et al. Biological application of quantum dots. Biomaterials. 2007;**28**:4717-4732

[20] Nabiev I et al. Fluorescent colloidal particles as a detection tools in biotechnology systems. In: Elissari A, editor. Colloidal Nanoparticles in Biotechnology. London-Singapore-NY: Wiley; 2008. pp. 133-168

[21] Malik P et al. Quantum dots for diagnosis of cancers. Advanced Materials Letters. 2013;**4**(11):811-822

[22] Chen F et al. Fluorescent CdSe/ZnS nanocristal-peptide conjugate for long-term, non-toxic imagimg and nuclear targeting in living cells. Nano Letters. 2004;**4**(10):1827-1832. DOI: 10.1021/nl049170q

[23] Alivisatos AP, Gu W, Larabell C. Quantum dots as cellular probes. Annual Review of Biomedical Engineering. 2005;**7**:55-76. DOI: 10.1146/annurev.bioeng.7.060804.100432. [PubMed]

[24] Raab RM, Stephanopoulos G. Dynamics of gene silencing by RNA interference. Biotechnology and Bioengineering. 2004;**88**:121-132

[25] Malik P et al. Computational Chemistry Laboratory, University of Delhi, Delhi 110 007, India. A review on CdSe quantum dots in sensing. Advanced Materials Letters. 2014;**5**(11):612-628

[26] Tiwari A et al. Intelligent Nanomaterials. USA: WILEY-Scrivener Publishing LLC; 2013. pp. 16-17

[27] Hongyou F et al. Surfactant-assisted synthesis of water-soluble and biocompatible semiconductor quantum dot micelles. Nano Letters. 2005

[28] Weal M et al. Emerging applications of fluorescent nanocrystals quantum dots for MMs detection. Proteomics. 2010;**10**(4):700-716

[29] Chomoucka J et al. Modern micro and nanoparticle-based imaging techniques. Sensors (Basel). 2012;**12**(11):14792-14820. DOI: 10.3390/s121114792

[30] Xu Y et al. Synthesis of CdSe/CdS core/shell quantum dots for biological sensing applications. Charlottesville, VA, USA: Department of Electrical and Computer Engineering, University of Virginia. 2006

[31] Shen L. Biocompatible polymer/quantum dots hybrid materials: Current status and future developments. 2011 Dec;**2**(4):355-372. Published online DOI: 10.3390/jfb2040355

[32] Alivisatos P. The use nanocrystals in biological detection. Nature Biotechnology. 2004;**22**:47-52

[33] Nann T. Phase transfer of CdSe–ZnS quantum dots using amphiphilic hyperbranched polyethylenimine. Chemical Communications. 2005;**13**:1735-1736

[34] Wang X-S et al. Surface passivation of luminescent colloidal quantum dots with poly(dimethylaminoethyl methacrylate) through a ligand exchange process. Journal of the American Chemical Society. 2004;**126**(25):7784-7785. DOI: 10.1021/ja0489339

[35] Rizvi BS et al. Semiconductor quantum dots as fluorescent probes for in vitro and in vivo biomolecular and cellular imaging. Nanotechnology Reviews. 2010;**1**(1-15). DOI: 10.3402/nano.v1i0.5161

[36] Mulder WJM et al. Quantum dots with a paramagnetic coating as a bimodal molecular imaging probe. Nano Letters. 2006;**6**(1):1-6

[37] Michalet X et al. Quantum dots for live cells, *in vivo* imaging, and diagnostics. Science. 2005;**307**(5709):538-544. DOI: 10.1126/science.1104274

[38] Lin Z et al. Methods for labeling quantum dots to biomolecules. Journal of Nanoscience and Nanotechnology. 2004;**4**:641-645

[39] Xing Y et al. Molecular profiling of single cancer cells and clinical tissue specimens with semiconductor quantum dots. International Journal of Nanomedicine. 2006;**1**(4):473-481

[40] Medintz I et al. Self-assembled nanoscale biosensors based on quantum dot FRET donors. Nature Materials. 2003;**2**:630-638

[41] Goldman ER et al. Conjugation of luminescent quantum dots with antibodies using an engineered adaptor protein to provide new reagents for fluoroimmunoassays. Analytical Chemistry. 2002;**74**:841-847

[42] Weal M et al. Emerging applications of fluorescent nanocrystals quantum dots for MMs detection. Proteomics. 2010;**10**(4):700-716

[43] Åkerman ME et al. Nanocrystal targeting *in vivo*. Proceedings of the National Academy of Sciences of the United States of America. 2002 Oct 1;**99**(20):12617-12621. Epub 2002 Sep 16

[44] Hardman R. Toxicological review of quantum dots: Toxicity depends on physicochemical and environmental factors. Environmental Health Perspectives. 2006;**114**(2):165-172

[45] Medintz IL et al. Quantum dot bioconjugates for imaging, labeling and sensing. Nature Materials. 2005 Jun;**4**(6):435-46. [PubMed]

[46] Vieira CH et al. Studying nanotoxic effects of CdTe quantum dots in *Trypanosoma cruzi*. Memórias do Instituto Oswaldo Cruz. 2011;**106**(2) ISSN: 0074-0276

[47] Jaiswal JK, Simon SM. Potentials and pitfalls of fluorescent quantum dots for biological imaging. Trends in Cell Biology. 2004;**14**:497-504

[48] Bhati, W. and Vishwa, A., Nanotechnology method comparison for early detection of cancer International Journal of Intelligent Systems and Applications. 2013;**3**:58-65. Published Online February 2013 in MECS

[49] Xiao L et al. Cytotoxicity of CdSe quantum dots and corresponding comparison with FITC in cell imaging efficiency. International Journal of Clinical and Experimental Medicine. 2017;**10**(1):753-759. ISSN:1940-5901/IJCEM0032047

[50] Derfus AM et al. Probing the cytotoxicity of semiconductor quantum dots. Nano Letters. 2004;**4**(1):11-18

[51] Kun M et al. Modeling distributed kinetics in isolated semiconductor quantum dots. Physical Review B 67, 125304 – Published 14 March 2003;**67**(12-15)

[52] Yong K-T et al. Imaging pancreatic cancer using bioconjugated InP quantum dots. ACS Nano. 2009;**3**(3):502-510

[53] Bharali DJ et al. Folate-receptor-mediated delivery of InP quantum dots for bioimaging using confocal and two-photon microscopy. Journal of the American Chemical Society. 2005;**127**:11364-11371

[54] Selvan ST. Silica-coated quantum dots and magnetic nanoparticles for bioimaging applications. Biointerphases. 2010;**5**(3):FA110-FA115

Chapter 3

Hydrogen Gas Detection by Mini-Raman Lidar

Tatsuo Shiina

Additional information is available at the end of the chapter

http://dx.doi.org/10.5772/intechopen.74630

Abstract

Now, Hydrogen gas is of particular interest as new energy source and dangerous material in nuclear facility. Fuel cell is started to use in home power generation system in 2008 and fuel cell vehicle (FCV) is commercialized from 2014 in Japan. On contrary, the Great East Japan Earthquake revealed the fear of hydrogen explosion on Fukushima Nuclear Power plant in 2011. Contact type hydrogen sensors induce changes on the gas flow, and the actual concentration cannot be known. It is hard to get the gas concentration distribution in hydrogen leakage area. We focused on optical remote sensing for the hydrogen detection. Raman scattering detection was accomplished for the hydrogen gas with a compact Diode Pumped Solid State (DPSS) laser-based Raman lidar. The quantitative measurement was conducted on the hydrogen gas concentration of 1 - 100% and the detectable distance of <50 m. Next, a LED-based mini Raman lidar was developed with the same optical design as the former one in viewpoints of its robust operation and usability in the nuclear facility. The high-speed photon counter was developed to follow the high repetition frequency of LED pulse of >500 kHz, and the quantitative measurements of hydrogen concentration were conducted on lab-experiment and at outdoor.

Keywords: hydrogen, lidar, Raman, scattering, LED

1. Introduction

Nowadays, there is a strong demand for remote sensing in near range. Needless to say, the fear of terrorism, the human daily lives whether at urban or at countryside have the threat of non-safety. The urban polluted air, smoke on chemical factory, dust from construction site pose a threat to human social activities. Certain dangerous gas detections in volcanic zone, construction site, and water inspection, air quality check in factory, hall, park, and so on, are desired for safe life. Certain gas/dust monitoring in construction site, and air quality estimation in daily life circumference are included in such demands for the improvement of working condition in factory and facility.

In the viewpoint of safety management, the working condition focus on the dust and leaked gas detections under construction site at outdoor. The mass media concerns to the atmospheric suspended materials represented as SPM (Suspended Particle Material), PM10/PM2.5, yellow sand, pollens, and also industrial dusts as dioxin, photochemical smog, and asbestos. The dust flow information, including the nuclear material, is strongly desired after the Great East Japan Earthquake. On indoor application, chemical gas flow in a factory, dust density measurement and dangerous gas monitoring such as CO, CO_2, CH_4, and H_2S in a tunnel construction site in mountain and under the sea are essential for such safety managements.

At the atmospheric monitoring indoor and outdoor, the "contact" sensors are utilized. These sensors are compact, cheap, and easy-to-use, while they need a lot to cover the whole area. A lot of sensors should be scattered in a certain area to get the target distribution and behavior. Low concentration gas and shooting out gas is difficult to perform the quantitative measurement [1, 2].

The word "LIDAR (Light Detection And Ranging)" is sometimes written as LiDAR for hard target detection and lidar for soft target detection. The former is utilized as a range finder to detect the distance up to the target. The latter is for monitoring the target condition and behavior at a certain distance. It is an ideal technique to monitor the suspended materials such as aerosol, dust, gas, and atmosphere remotely and safely [3–5]. Unlike a transmissometer which consists of an separated pair of transmitter and receiver, lidar can fix at a point and scan the observing direction. Then it can cover a certain area remotely. Lidar can obtain the distribution and change of the target in monitoring site by scanning the observation direction. Its observation range is enough long of a few to a hundred kilometers. In contrary, recently it applied for near range monitoring within a few hundred meters. Its application is industrial use.

Lidar technique utilizes various kinds of scattering. Mie scattering, Rayleigh scattering, Raman scattering, and so on. Mie scattering is generated by larger materials such as aerosols and dusts, while the Rayleigh scattering is caused by smaller materials such as molecules configuring the atmosphere. Raman scattering is brought about every material but it shifts wavelength of scattering light from the incident light one. It is convenient for target gas detection.

In this study, we developed the compact Raman lidar for hydrogen gas detection. To prevent the explosion accident of hydrogen gas, the sensor should detect the target gas of the low concentration of 1/4 of the explosion limit. In the case of hydrogen gas, its concentration is 1% [6–8]. The optical intensity of the lidar transmitter is designed to be safe for human eye and skin. This study has been started to apply for leaked gas detection in a hydrogen gas station [9, 10]. After that it has expanded to adapt to the nuclear facility. This lidar should be compact and easy-to-use. It is also applied to monitor the air qualities of working environment including hydrogen gas, atmospheric nitrogen gas and water vapor.

This report states the developments of the Diode Pumped Solid State (DPSS) laser-based compact Raman lidar and the LED-based mini-Raman lidar. At first Raman scattering and lidar principle are briefly mentioned, and after that, the development of concrete Raman lidars is explained. Some results of quantitative measurements are shown with actual observation data.

2. Hydrogen generation – controlled and unprepared

Utilization of hydrogen energy is undergoing, which is led mainly by the Agency for Natural Resources and Energy in the Ministry of Economy, Trade and Industry, Japan to promote the spread of fuel cell car (FCV) and home-use fuel cell system.

Fuel cell generates electricity by chemical reaction between hydrogen and oxygen. It causes only water vapor and its power generation efficiency exceeds 30%. Hydrogen gas is stored in a high-pressure tank at 35 MPa in hydrogen station and FCV. 96 hydrogen stations were built in 2016 in Japan, and it will increase 160 stations in 2020. FCV were installed 2000 cars in 2016, and bus and forklift truck types were installed into on-site, too. The agency aims to expand it up to 40,000 cars in 2020. Now, FCV seems to conceal itself under electric vehicle (EV). FCV, however, can drive for a long travel distance of 700–850 km at one-time insufflations of hydrogen of 3 min [11]. For that reason, EV is good for small automobile for short distance, while FCV has a merit for large-size automobile for long distance.

Home-use fuel cell system is established its technologies in complex with cogeneration. Hydrogen is generated from urban gas or LP gas, and chemically reacts with oxygen, then generates the electricity. Exhaust heat on electricity generation is utilized to the other purpose (for example hot water supply). 200,000 systems have been already installed up to 2016. Reducing the price of the system is undergoing.

Hydrogen gas on the nuclear reactor is generated by the different process against the above. In general, the nuclear fuel rod boils water with its high temperature in nuclear reaction. At the Great East Japan Earthquake, it is considered that the tragic event was caused by the following process. The nuclear reactor was stopped automatically, while the fuel cell rod continued heating. Because the cooling system was fallen down, the water level was lowered and the rod was naked. Although the nuclear fuel rod was covered with zirconium, it reacted with water on the high temperature. As a result, zirconium took oxygen away from the water, the hydrogen gas was generated.

Under the controlled condition to generate the electricity or the unforeseen circumstances, hydrogen gas monitoring is essential and should prevent the accident.

3. Optical sensing for hydrogen gas

Hydrogen is flammable gas. It explodes at the concentration of >4% under the oxygen circumference [12]. The Fire Defense Agency defined that the gas sensor must have the minimal detection of the one fourth of the explosion limit of the target gas. It is essential to detect such flammable gas before the risk of its explosion. Methane, Carbon oxide, and so on have absorption lines in the near infrared spectrum, and the differential absorption technique can be adapted for these gas detections. The hydrogen gas, however, has its absorption line in < 110 nm and does not have any clear absorption lines in ultra violet, visible – near infrared spectra [13]. The hydrogen gas detection is only conducted by contact chemical sensors such as combustible material, catalyst materials (Pt or Pd), and semiconductor type (SnO_2) [1, 2]. These sensors are high-sensitive to detect its concentration of a few tens ppm, while it is hard to get the gas distribution in whole of monitoring site. Furthermore, the contact sensor disturbs the gas flow, and its sensing direction influences to the measurement value. Its response time is slow, too.

Hydrogen gas has strong Raman effect, which depends on material. Raman scattering shifts the wavelength of the scattering light from that of the incident light. In general, the intensity of Raman scattering is 3 digits smaller than that of the Mie scattering [14, 15]. When electrons in atmospheric molecules (or certain gas, atom, and so on) are resonant with photons, most photons are elastically scattered (it is Rayleigh scattering), such that the scattered photons have the same optical frequency and wavelength as the incident light (photons). A small fraction of such photons is scattered with a frequency different from the original frequency of the incident photons. The frequency shift is usually lower than and partially higher than the original; Stokes scattering as longer wavelength and anti-Stokes scattering as shorter wavelength. The interaction of light with a material in a linear region allows absorption and emission of a photon precisely matching the difference in energy levels of the interacting electron. The energy absorption of the molecules causes Stokes scattering, while the energy loss of the molecules generates the anti-Stokes scattering. Stokes scattering shifts the wavelength of the incident light to the longer side. Anti-Stokes scattering shifts that to the shorter side as shown in **Figure 1**. Intensity of Anti-Stokes scattering is smaller than that of Stokes scattering. In general, this trend is remarkably strong when the Raman shift becomes large. In this study, Stokes scattering of hydrogen gas is selected for the lidar detection target under these backgrounds.

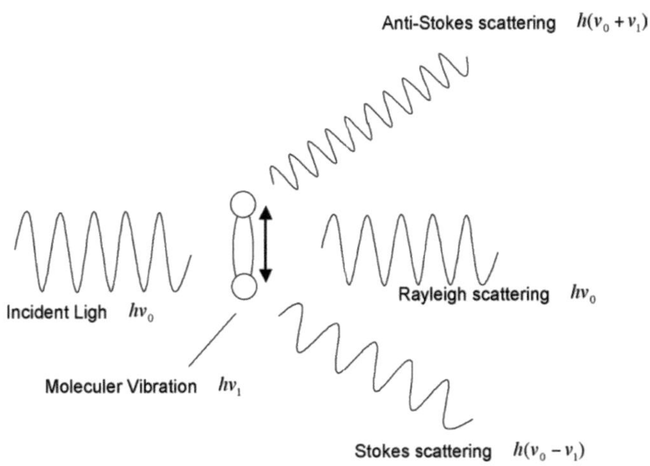

Figure 1. Raman scattering.

Target	Raman shift (cm^{-1})	Wavelength (nm)
Nitrogen	2331	380
Water vapor	3652	400
Oxygen	1556	369
Hydrogen	4160	408

Table 1. Raman shift and its wavelength against 349 nm light source.

To estimate the quantitative concentration of hydrogen gas, Raman scattering intensities of atmospheric nitrogen is utilized. When the concentration of hydrogen gas is negligible amount than that of atmospheric nitrogen, the ratio between the Raman scattering intensities of hydrogen gas and atmospheric nitrogen can indicate the concentration of the hydrogen gas quantitatively. The ratio is calibrated by the relative Raman scattering cross sections of hydrogen and nitrogen, which is about 3:1 [5]. **Table 1** shows the concrete Raman shifts and their shifted wavelengths against the 349 nm laser light source.

4. Compact Raman lidar

The principle of lidar is illustrated in **Figure 2**. The pulsed laser beam was collimated and fired into the atmosphere with almost parallel to the receiver's optical axis. The optical receiver is adjusted its field of view (FOV) to detect the lidar echo from the target with an adequate signal-to-noise ratio. The receiver's FOV decides the detecting range and the blind area, which the transmitting beam cannot enter within a receiver's FOV, especially in near range. It also affects the detecting intensity of the background light. As a traditional lidar has individual optics (biaxial optics) for a laser transmitter and a receiver, the system has a blind area. The blind area also causes in the case of coaxial optics that the beam is transmitted within a receiver's area. As FOV is wider, or the beam divergence is wider, the blind area will be shortened, while the background light enters a lot into the receiver. As a result, the signal-to-noise ratio of the lidar echo will be worse.

The lidar echo is detected by high sensitive detectors such as photomultiplier and Avalanche Photo Diode (APD), and stored in PC via digitizer. When the lidar echo becomes weak and hard to detect an analogue signal, photon counting method is selectable.

The compact lidar is suitable to the near range observation of less than a few hundred meters, while the traditional large lidar is good at the long-range observation from a few kilometers to a hundred kilometers. The compact lidar, however, is not a down-size one of the large lidar. **Table 2** shows the spatial and time scales in atmospheric phenomena. The large phenomena such as HP/LP have a large spatial and time scales, while the small phenomena like tornado takes small spatial and time scales. That is, the small phenomenon becomes small structure and quick motion. To detect and visualize such small phenomena, the compact lidar should follow to its quick motion with high resolution. Furthermore, near range detection is often a sensing in human living space. In that meaning, the transmitting optical power should be

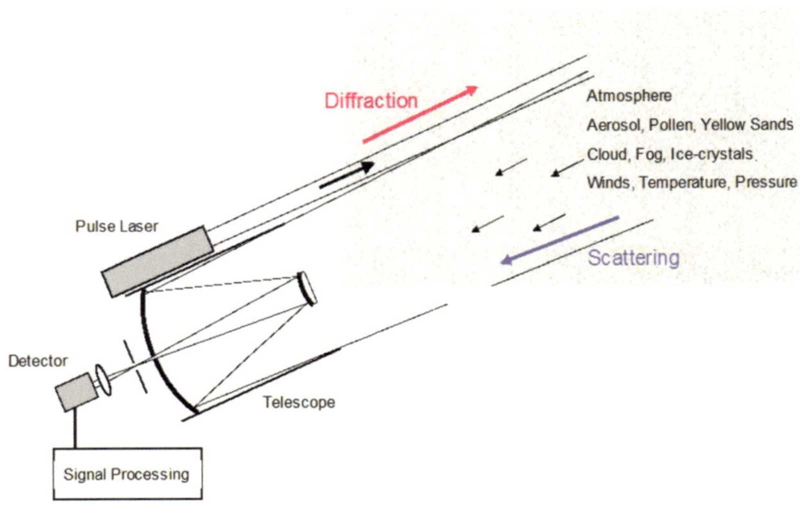

Figure 2. Lidar principle.

lower for eye-safety. It causes the limit of transmitting power and the selection of optical wavelength.

The Raman scattering intensity is estimated by Eq. (1).

$$P_R = \frac{\lambda_0}{\lambda_R} P_0 \chi \rho \sigma_R L \eta_D \tag{1}$$

where P_0, P_R are wavelength λ_0, λ_R of laser light source and Raman scattering intensity, respectively. χ is molar function of the target, ρ is number density of the molecules, σ_R is Raman scattering cross-section, L is pulse length and η_D is receiver's optical efficiency. The quantitative

Phenomena	Spatial scale	Time scale
HP/LP	1000 km	10 h
Typhoon	100 km	3 h
Convection	50 km	2 h
Thunder clouds	10 km	1 h
Cumulus	2 km	10 min
Down burst	600 m	7 min
Tornado	200 m	5 min
Boundary layer	60 m	10 s

Table 2. Spatial and time scales of atmosphere.

http://dx.doi.org/10.5772/intechopen.74630

measurement is calculated by the intensity ratio of nitrogen echo and hydrogen echo. The lidar echo is inversely proportional to the distance to the power of square. But the above intensity ratio cancels the intensity decrease [6].

The concept of the hydrogen remote sensing is to install the "remote sensor" into the area monitoring in a hydrogen station. The sensor is put on a small car and monitors the inside area of the station on the street. In that meaning, the remote sensor should be compact. The eye-safety should be considered, too.

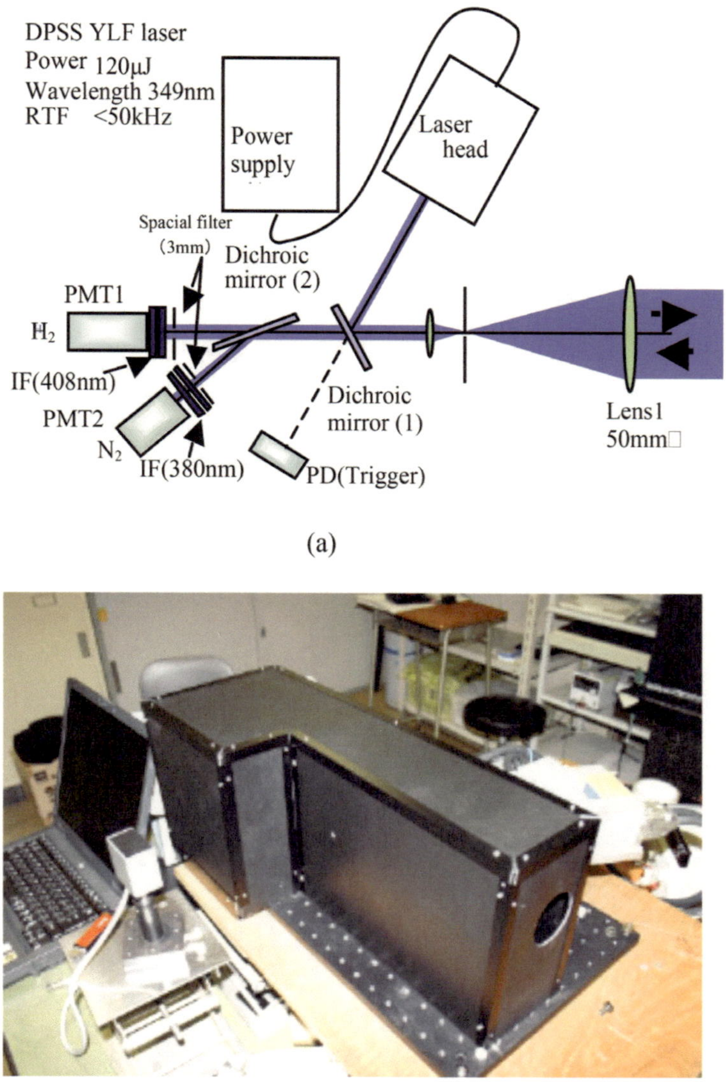

Figure 3. Optics of compact Raman lidar. (a) Compact Raman lidar—optical setup. (b) Outlook of compact Raman lidar.

In this study, we have developed the "in-line" type compact Raman lidar [16–21]. The in-line type optics has no blind area in principle. The in-line lidar has common optics for a transmitter and a receiver. An optical divider, here dichroic mirror, to separate the receiving echo and transmitting beam is installed. We designed and developed the compact Raman lidar to detect leaked hydrogen gas as minimum setup. It means that the system should be compact, light-weight and low electric power consumption. Our goal is the detection of the hydrogen gas concentration of 1% within the observation range of 0–50 m. The lidar optics is illustrated in **Figure 3**. The DPSS laser is utilized for lidar transmitting beam. Its power is 120 μJ and the wavelength is 349 nm. The receiver's aperture is 50 mmφ. The transmitting beam is expanded to 10 mmφ to be safe for human eye even in front of the system. The wavelength of the incident beam is decided to obtain the strong Raman signal. Shorter wavelength causes the larger Raman signal than the longer wavelength. The receiver's FOV was 2 mrad. The Raman shifted echoes passes through the first dichroic mirror (1), which is a divider, and echoes never returns to the light source. The Raman echoes are divided by the second dichroic mirror (2) depending on the wavelength for each Raman detection. The detectors are PMTs; one is for hydrogen gas (408 nm) and the other is for atmospheric nitrogen gas (380 nm). The interference filters of the wavelength widths of 3 nm are fixed on their PMTs. The nitrogen signal is used to estimate the quantitative concentration of hydrogen gas. The system size is 58 × 36 × 23 cm. It can put on the small desk.

Figure 4 shows the detected Raman echo of atmospheric nitrogen. Its signal was obtained from 1.5 to 50 m. The nearest echo in 0–1.5 m in **Figure 4(a)** was spoiled because of the misalignment and the aberration of the lenses. **Figure 4(b)** shows the range-corrected nitrogen Raman echo, which is corrected by the square of the distance and the overlap function between the transmitter and the receiver. The corrected echo intensity becomes flat and the fluctuation is minimum at any point within the observation range. The constant intensity indicates stable concentration of atmospheric nitrogen. This nitrogen echo is utilized to estimate the concentration of hydrogen gas in the observation range by taking the ratio between them.

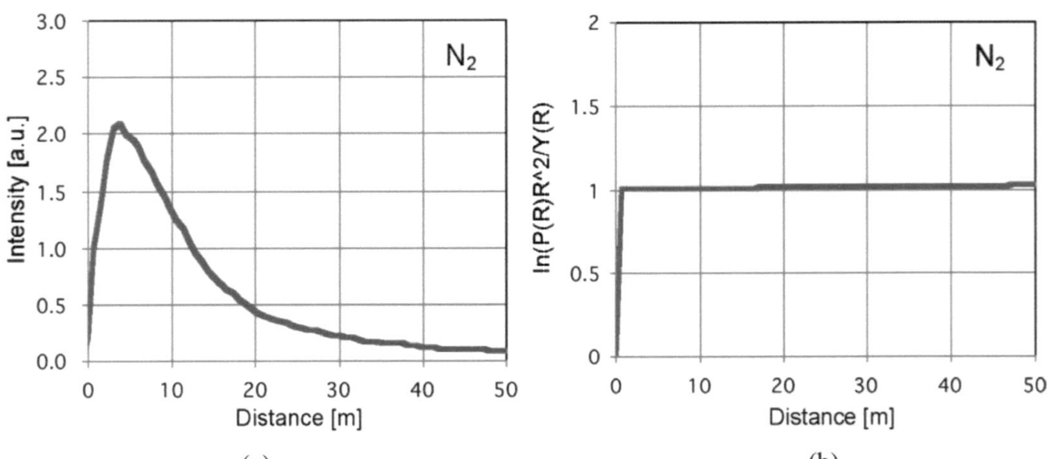

Figure 4. Lidar echo of atmospheric nitrogen gas. (a) A-scope signal. (b) Range-corrected data.

A quantitative measurement of the Raman scattering echo intensity from hydrogen gas was conducted. The hydrogen gas was filled in a gas cell of the size of 10 cm cube, of which window is quartz. The cell was positioned at a distance of 7–8 m from the lidar system and put with a small tilt against the optical axis to reduce the reflection and fluorescent lights to return the lidar. Although the reflection and fluorescence lights appear on the hydrogen and nitrogen echoes, they can be subtracted out. Furthermore, the Raman echo from hydrogen gas fluctuates in actual gas leaks, and it can be separated from the reflection and fluorescence lights from a hard target, which are constant in time. The observation distance of 8 m is defined as the safe distance for the explosion protection of hydrogen gas. The lidar echo was measured by an oscilloscope with a photomultiplier tube (PMT) operating in analogue mode.

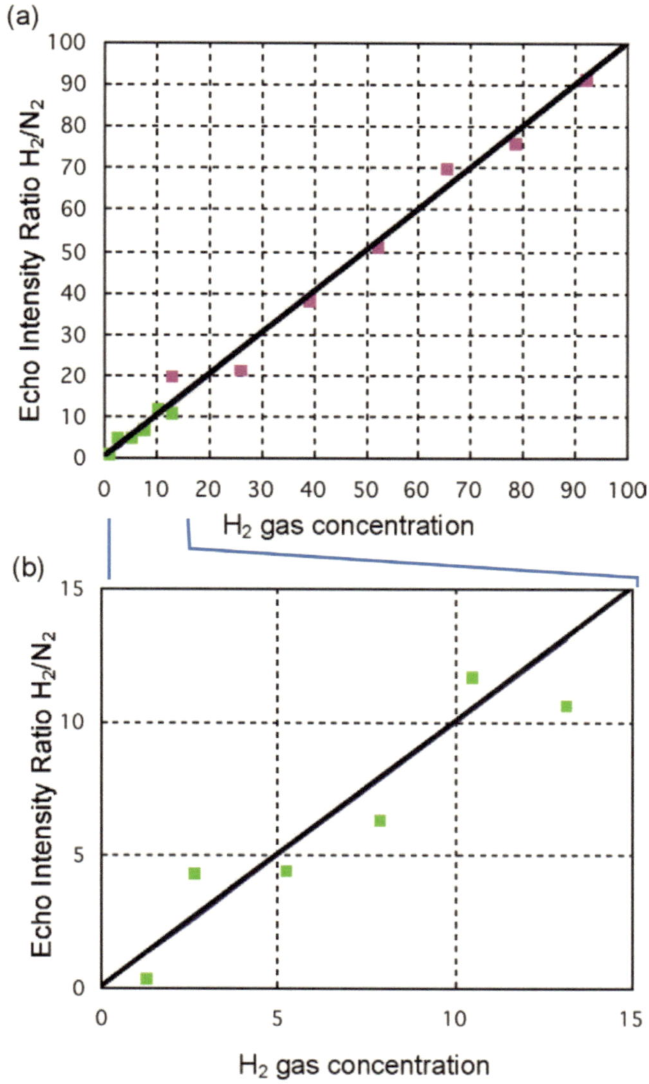

Figure 5. Estimation of hydrogen gas concentration. (a) High concentration of about 10–100% and (b) low concentration of 0–15%.

The echo intensity was obtained by the developed Raman lidar system. The concentration of hydrogen gas was controlled by adjusting the gas pressure in the cell. The results are shown in **Figure 5**, for (a) high concentration of about 10–100% and (b) enlarge graph of low concentration of 0–15%. These results were calibrated by subtracting the reference signal of the empty gas cell. The summation was about 30,000 shots. It is equal to a measurement time of about 30 s. The echo intensity ratio is linearly proportional to the hydrogen gas concentration. The result indicates the successful estimation down to the low concentration of 1%. The estimation error becomes less than ±1%. The fluctuation in the echo intensity was mainly due to the background light, unstable power of the laser beam and thermal noise of the detector. The actual leaked gas fluctuates more drastically.

5. LED-based mini-Raman lidar

The developed DPSS-based compact Raman lidar is designed to be eye-safe, while it is hard to install it into a certain monitoring site. The laser beam is possible to focus on a point, and it is not acceptable especially in dangerous gas environment because of its ignition. It is also hard to use it at outdoor with no AC power supply. In such demands, we started to develop the LED-based mini-Raman lidar with low optical power.

At first, we had the experiment to obtain the minimum power to detect the Raman echo of hydrogen gas as shown in **Figure 6**. The laser beam is entered into the quartz-windows cell filled with hydrogen gas. The hydrogen gas concentration was 100%. The Raman scattering echo was detected at the vertical direction. It is because the Raman intensity becomes maximum at the perpendicular direction against the optical axis of the transmitting beam. The

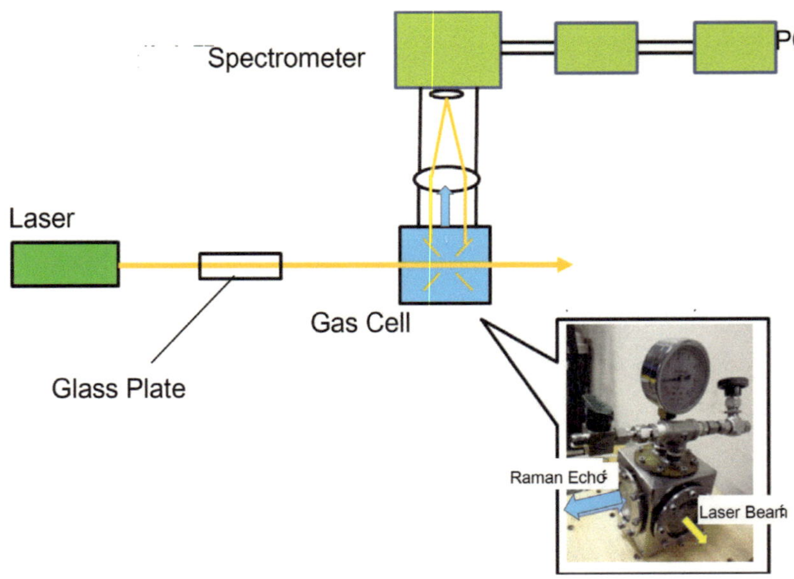

Figure 6. Verification of Raman scattering echo measurement with the minimum light intensity.

laser power was gradually decreased up to the Raman detection limit, which is estimated 1 W (=10 nJ/10 ns) by the numerical analysis. The reflection light by the glass plate was utilized to decrease the laser intensity efficiently and stably. The experimental result is shown in **Figure 7**. It shows that the Raman scattering echo was successfully detected by the estimated optical power of the transmitting beam.

Considering the above results, we developed the LED-based mini-Raman lidar. The lidar optics is biaxial type and the gas detection method is as same as the laser based compact Raman lidar as shown in **Figure 3**. The LED lamp module of the optical power of 1 W was used as the lidar light source. The wavelength of LED light is 365 nm, and the wavelengths of Raman scattering echoes become 400 nm for atmospheric nitrogen, 420 nm for water vapor and 430 nm for hydrogen gas with the same Raman shift shown in **Table 1**. The LED mini-Raman lidar optics is shown with photo in **Figure 8**. The optical setup is the same as the DPSS-laser based compact Raman lidar. The LED-based Raman lidar added the water vapor port by changing the optical interference filter with a filter wheel. All of the wavelength widths of the interference filters were 10 nm. It is the same width the used LED spectrum.

Although LED beam is hard to collimate, it never focuses after the transmission. It gives advantage in the closed space in the viewpoint of human safety concern. In this study, the transmitting beam was enlarged to 50 mmφ. Its divergence was about 10 mrad. The LED pulsed power was <1 W (=10 nJ/10 ns), while the pulse repetition frequency was 500 kHz. The receiving telescope was Cassegrain type of the aperture of 127 mmφ. The receiver's FOV was 3 mrad. The transmitting power is so weak, and the echo detection is photon counting with photomultipliers.

The system is so compact and handy that one can use it in any observation direction. The system power consumption is about 2 W and can operate with a small PC battery in 8–10 h. The observation range is aimed up to 20–30 m for hydrogen Raman signal. This observation range design comes from the facility space of a nuclear reactor to detect the hydrogen gas leakage. The specifications are summarized in **Table 3**.

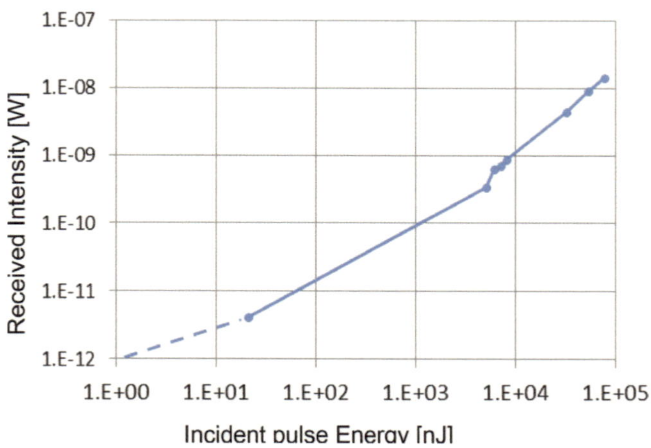

Figure 7. Raman scattering echo detection with lowered incident power.

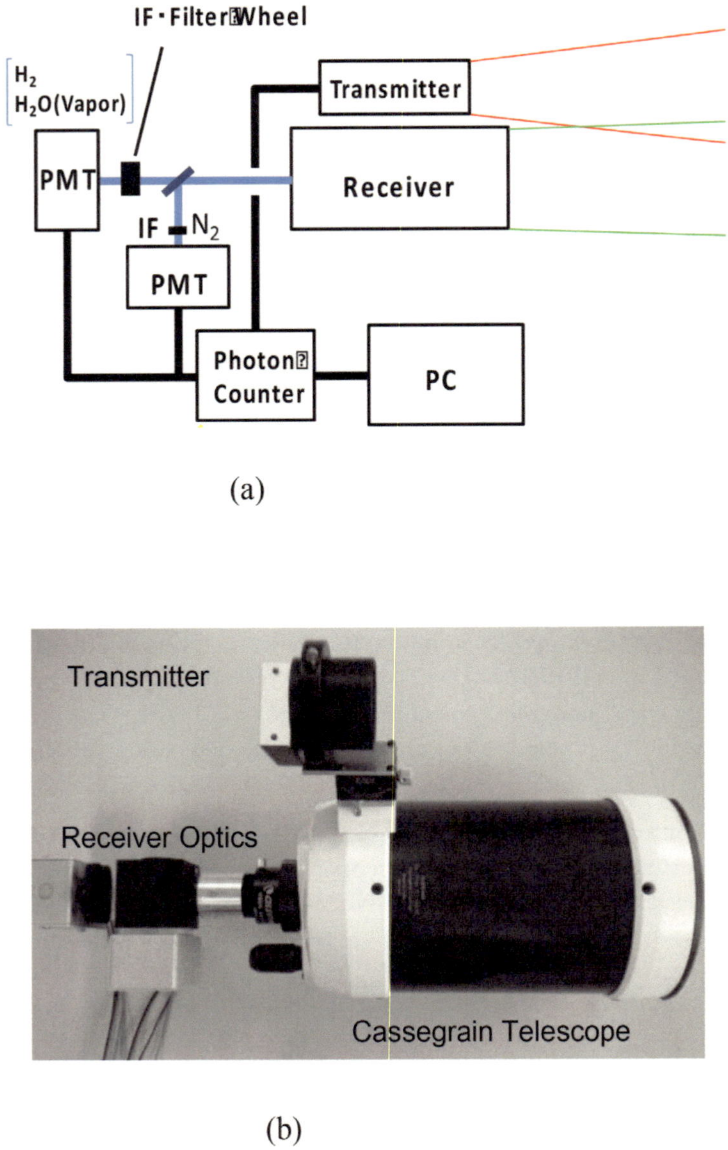

Figure 8. LED mini-Raman lidar. (a) Optical structure. (b) Lidar setup.

The high-speed photon counter was developed, too. It can follow the pulse repetition frequency of >500 kHz. Its time resolution is 1 ns. It is equivalent to the spatial (distance) resolution of 0.15 m. It has 2 channels, and their simultaneous measurements are possible with a single input or output trigger signal.

Figure 9 shows the observed Raman echo of the atmospheric nitrogen gas and the water vapor. The observation direction is nearly horizontal. The experiment was conducted in night time. Trees were captured at 80 m ahead. Accumulation was 3.5 min. Nitrogen echo was

Transmitter	
LED Pulse Power	750mW
Wavelength	365nm
Beam Diameter	50mmϕ
Beam Divergence	10mrad.
Repetition Frequency	>450kHz
Receiver	
Telescoe	Cassegrain
Diameter	127mmϕ
Field of View	<5mrad.
Interferene Filter	$\Delta \lambda$ =10nm
Detector	Photomultiplier

Table 3. Specifications of LED mini-Raman lidar.

obtained up to 70 m (**Figure 9(a)**). After that the fluorescence echo of the tree was also detected at 80 m. The atmospheric nitrogen is stable and its echo waveform becomes smooth. On the other hand, the water vapor echo had fluctuation because of low counts (**Figure 9(b)**). It was drastically changed by depending on the humidity. In the figure, the humidity was 25% on Feb. 7 in fine day, while it was 90% on Feb. 9 in rain day. The observation range of the water vapor was 70 m, which is equal to that of the nitrogen echo. The fluorescence echoes of trees were detected, too.

The quantitative measurement of hydrogen gas was conducted with a large gas chamber as shown in **Figure 10**. The hydrogen gas chamber has an aperture of 200 mm and a length of 500 mm. Pyrex glass of 10 mm thickness was used for light path windows to restrict the fluorescent light. It was set 5 m ahead from the LED mini-Raman lidar. The transmitting beam was adjusted its optical path against the receiver's field of view to cross both of optical paths at the chamber.

The hydrogen gas concentration was adjusted by vacuuming the air inside the chamber of 0.5 Pa and inserting the 100% hydrogen. This procedure repeated twice, and the hydrogen gas concentration of 75% was prepared. We also confirmed the air and hydrogen gases were well mixed in the chamber. After the first measurement, hydrogen gas concentration set to half, that is, 37.5% by vacuuming the inside gas of 0.5 Pa and inserting the air. Repeating these procedures, hydrogen gas concentration was adjusted as 18.8%, 9.4%, 4.7%, and 2.4%.

Experimental results are shown in **Figure 11**. The accumulation time was 3.5 min. The hydrogen Raman echo was appeared on the fluorescent light from the Pyrex glass. **Figure 11** shows enlargement of the peak change of the echo signals. Depending on the hydrogen gas concentration, the peak counts were changed. The echo counts of hydrogen gas Raman signal were subtracted from the reference echo (0% hydrogen gas). The results were summarized in **Figure 12**. In **Figure 12(a)**, hydrogen Raman echo counts at each detected distance were

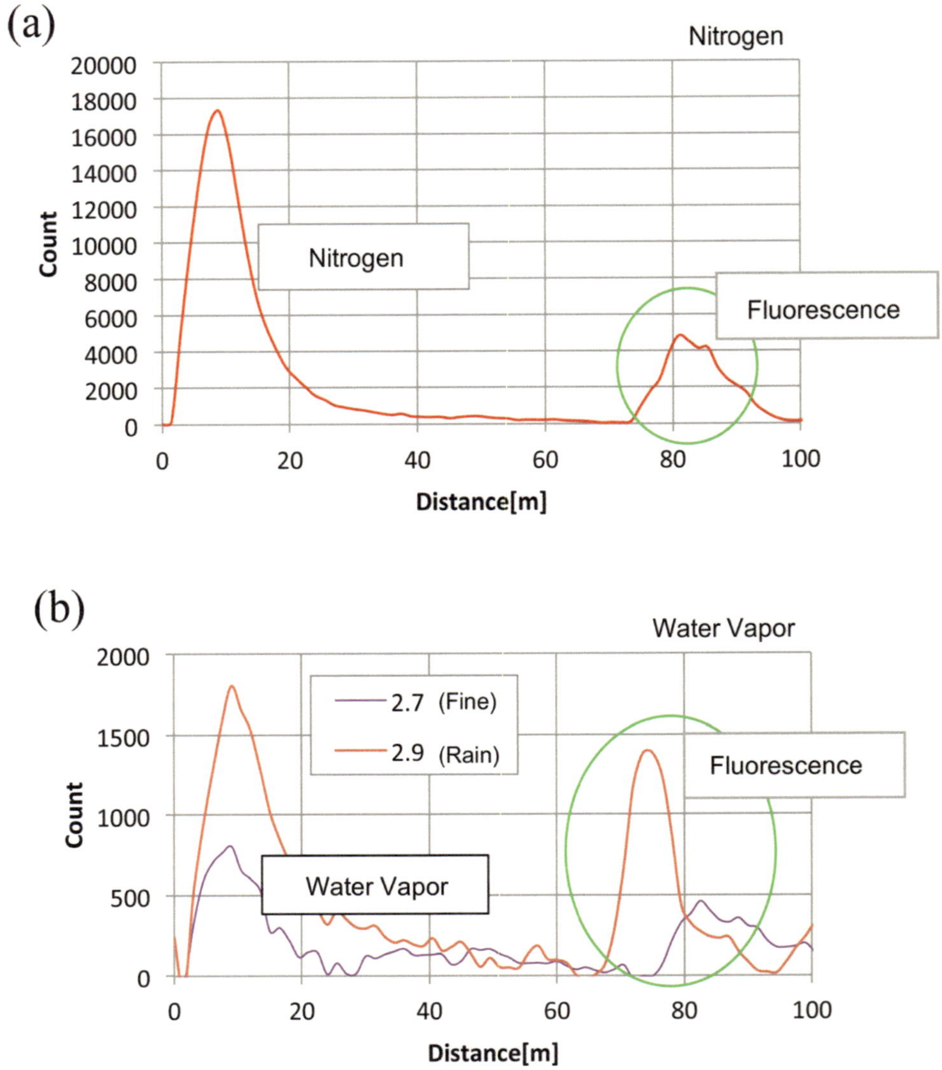

Figure 9. Atmospheric measurement by LED mini-Raman lidar. (a) Nitrogen Raman echo (b) Water Vapor echo.

graphed. The Raman echo counts at 5.25 and 5.4 m had the higher slant, while the firster (initial distance arising echo) and the later parts (main part of fluorescent light) of the peak signals showed lower slant. It means that the Raman scattering appears firster than the fluorescent light. The lidar counts at each distance in **Figure 12(a)** was summed as shown in **Figure 12(b)**. The errors were large, which came from the inexact concentration control of the hydrogen gas and the fluctuation of the fluorescent light, while the linear change was obtained down to the lowest concentration of a few percent. The Raman echo counts were remained at the 0% hydrogen gas concentration. We regarded hydrogen concentration after the procedure of vacuuming the content gas of 0.5 Pa and inserting the air two times as equal to 0%. It caused the remained echo counts at 0% concentration. And the fluctuation of the fluorescent light peak echo will cause it, too.

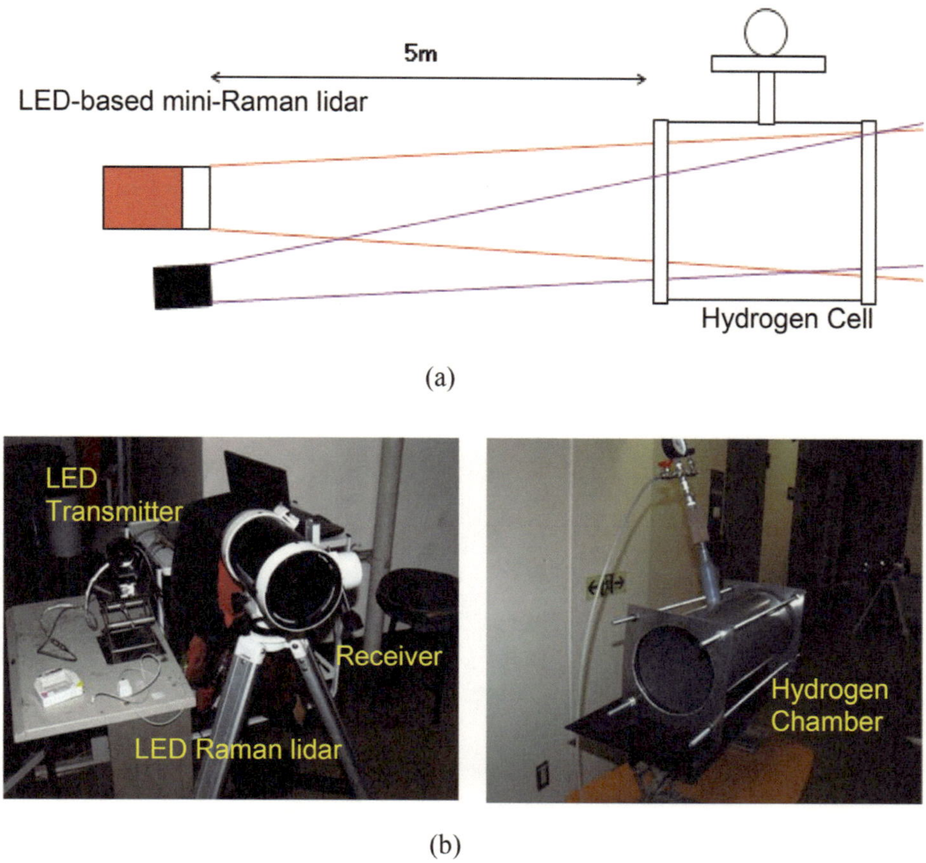

Figure 10. Hydrogen gas quantitative measurement. (a) Quantitative measurement of hydrogen gas. (b) System setup and arrangement.

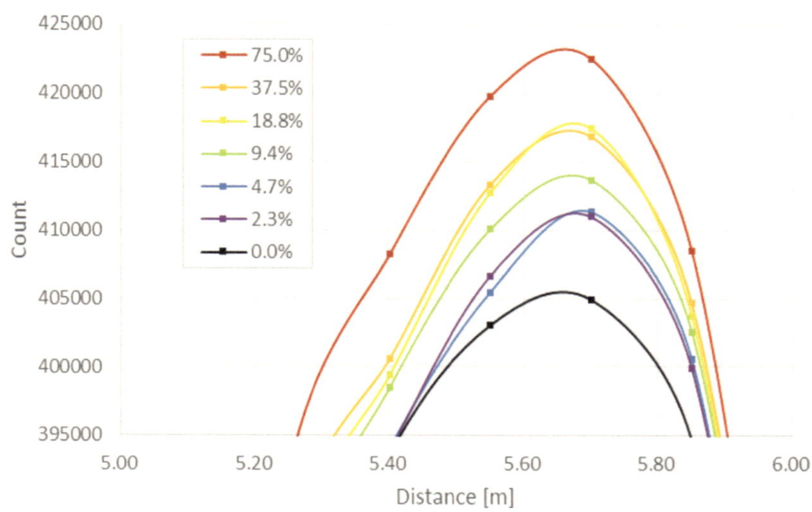

Figure 11. Peak signal changes depending on hydrogen gas concentration.

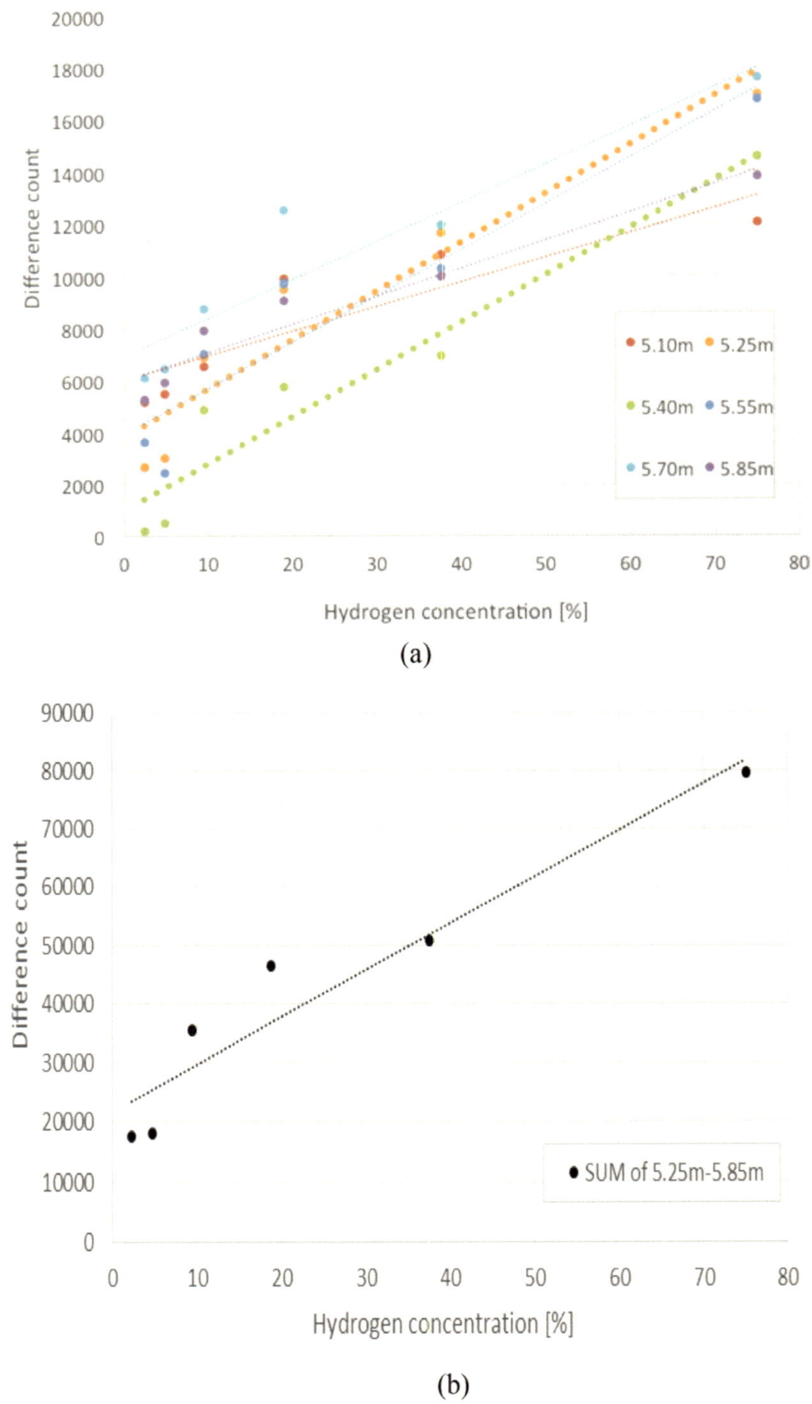

Figure 12. Change of hydrogen Raman echo counts depending on hydrogen gas concentration. (a) Echo changes at each detected distance. (b) Count sum at each concentration.

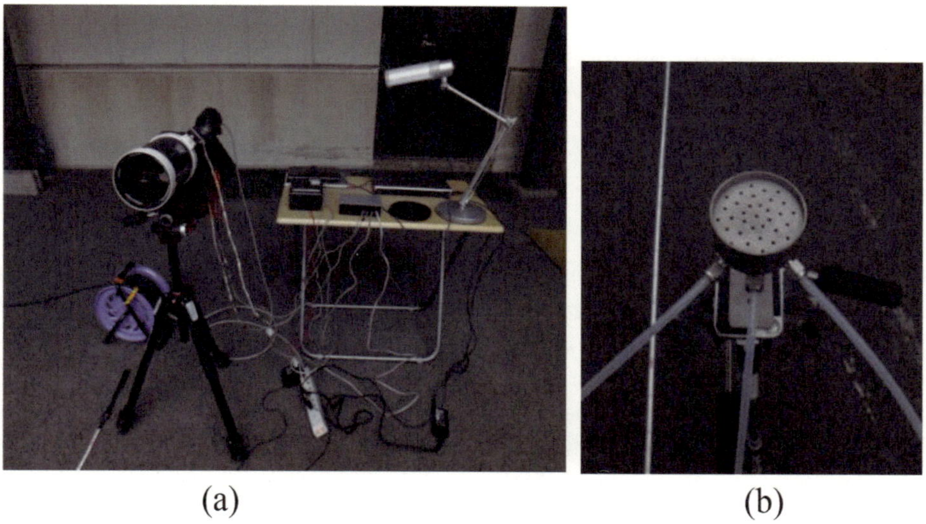

Figure 13. Hydrogen leakage measurement experiment outside. (a) LED mini-Raman lidar setup. (b) Hydrogen nozzle.

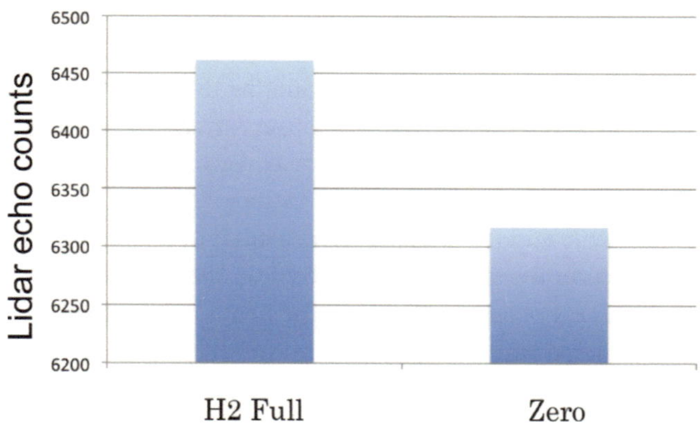

Figure 14. Measurement result of hydrogen leakage gas outside.

The hydrogen leakage experiment was conducted at outdoor in night time as shown in **Figure 13**. The hydrogen leakage nozzle shown in **Figure 13(b)** was fixed at 2.5 m ahead from the mini-lidar. The optical alignment between the transmitter and the receiver was adjusted to overlap fully. The leaked hydrogen gas was 50 l/min. The result is summarized in **Figure 14**. The accumulation time was 3.5 min. It is the result of 3 average times. The obvious difference between the hydrogen gas leakage and leakage stop conditions was obtained. The detected counts on the leakage stop condition were caused by the fluorescent light from the materials near the hydrogen nozzle. As compared with the lab experiment in **Figure 12**, the hydrogen Raman counts was low. The leakage gas volume, especially gas flow width against the optical axis was shorter than the hydrogen gas chamber or the optical pulse length. As a result, the counts rate outdoor was smaller than the lab-experiment. After that, the authors repeated the quantitative measurement with more large chamber at the outdoor field. The same result in the lab-experiment was obtained.

6. Discussions

In the lab-experiment, that is, the hydrogen cell/chamber used experiments at indoor, the linear change of the lidar counts was obtained depending on the hydrogen concentration up to 1%. The accuracy is enough to evaluate the concentration quantitatively. The DPSS laser-based lidar can perform it within a few seconds by adequate setup. The LED-based lidar needs more time to evaluate the concentration. The evaluation error is remained on the current setup. At outdoor experiment, the obvious hydrogen Raman echo by the LED Raman lidar was detected at the gas leakage situation, while it is not easy to verify the concrete concentration. It is not only the system task, but also the gas flow fluctuation. As the leakage area was narrower than the pulsed beam length, the counts rate became lower. The accumulation time, beam size, and transmitting power intensity should be balanced to satisfy the demands.

The hydrogen gas detection is not only to estimate the precise concentration, but also on the easy-to-use and easy-to-install at on-site. The LED-based mini-Raman lidar is designed under this consideration. Its gas detection optics is as same as the laser-based Raman lidar, while the LED module is installed into the system. The observation range becomes short, while the system becomes so small, light-weight, low power consumption, and safe. The power consumption is enough small to operate with a small buttery. The LED-based mini-Raman lidar needs the accumulation time, and it is adequate to install it to the regular observation fixed on a certain position and monitor the stored gas on a ceiling. When Ultra Violet/Near Ultra Violet Laser Diode (UV/NUV-LD) is possible to utilize, the beam collimation will be improved and it can increase signal-to-noise ratio with the same optical pulse power. It will shorten the accumulation time, and become one of the solutions to detect the leakage hydrogen gas in the emergency case under no electrical power.

7. Conclusions

We developed the compact and the mini-Raman lidars for quantitative measurement of hydrogen gas. The current minimal detection limit of hydrogen gas concentration is 1% at the observation range of 0–50 m with DPSS-based compact lidar and 0–20 m with LED-based mini-lidar. The accumulation time is 30 s for the DPSS laser-based compact Raman lidar and 3.5 min for the LED-based mini-Raman lidar.

Now, the DPSS laser-based compact Raman lidar was improved to detect the ionization effect by nuclear materials. On contrary, the LED-based mini-Raman lidar expanded the target to the environmental measurement. The Raman scattering signal is a proof of a target gas existence. Each gas has its own Raman shift, and the optical sensing method of Raman lidar becomes one of solutions in the actual atmosphere, which contains complex materials.

The Raman signal is weak. It is 1/1000 of the Mie scattering signals. Our current experiments were limited in nighttime. When the DPSS laser-based lidar is fixed in a sunshade area, it will act with enough accuracy. At indoor, the artificial light was no effect for Raman signal detecting by excluding for its direct insertion to the detector even if in the case of the LED-based lidar.

The hydrogen has a potential to become major energy resource. It will be popular in daily life near future. On contrary, its dangerousness is important to recognize once more. Remote sensing device for leaked hydrogen gas is essential. Compact and mini-Raman lidars will be one of the solutions.

Acknowledgements

This study was granted by Chubu Electric Power Co., Inc. from 2013 to 2016. Shikoku research institution Inc. was cooperated on the outdoor experiment, too. The author really appreciates to these companies.

Author details

Tatsuo Shiina

Address all correspondence to: shiina@faculty.chiba-u.jp

Graduate School of Engineering, Chiba University, Yayoi-cho, Inage-ku, Chiba-shi, Japan

References

[1] Gu H, Wang Z, Hu Y. Hydrogen gas sensors based on semiconductor oxide nanostructures. Sensors. 2012;**12**:5517-5550

[2] Zeng XQ, Latimer ML, Xiao ZL, Panuganti S, Welp U, Kwok WK, Xu T. Hydrogen gas sensing with networks of ultrasmall palladium nanowires formed on filtration membranes. Nano Letters. 2011;**11**:262-268

[3] Measures RM. Laser Remote Sensing: Fundamentals and Applications. New York: John Wiley & Sons; 1984

[4] Fujii T, Fukuchi T, editors. Laser Remote Sensing. Taylor & Francis; 2005

[5] Weitkamp C, editor. Lidar: Range-Resolved Optical Remote Sensing of the Atmosphere. Springer; 2005

[6] Ninomiya H, Yaeshima S, Ichikawa K, Fukuchi T. Raman lidar system for hydrogen gas detection. Optical Engineering. 2007;**46**:094301

[7] Ninomiya H, Asahi I, Sugimoto S, Shimamoto Y. Development of remote sensing technology for hydrogen gas concentration measurement using Raman scattering effect. IEEJ Transactions on Electronics, Information and Systems. 2009;**129**:1181-1185

[8] Thomas CE. Fuel cell and battery electric vehicles compared. Hydrogen Energy. 2009; **34**:6005-6020

[9] Asahi, Sugimoto S, Ninomiya H, Fukuchi T, Shiina T. Remote sensing of hydrogen gas concentration distribution by Raman lidar. Proceedings of SPIE. 2012;**8526**:85260X

[10] Shiina T. Optical design for near range lidar. Proceedings of SPIE. 2010;**7860**:78600B

[11] Suppes G. Plug-in hybrid with fuel cell battery charger. Hydrogen Energy. 2005;**30**:113-121

[12] Basquin S, Smith K, editors. Hydrogen Gas Safety Self-Study. Los Alamos National Laboratory. 2000

[13] Backx C, Wight GR, Van der Wiel MJ. Oscillator strengths (10-70eV) for absorption, ionization and dissociation in H_2, HD and D_2, obtained by an electron-ion coincidence method. Journal of Physics B: Atomic and Molecular Physics. 1976;**9**:315-333

[14] Leonard DA. Observation of Raman scattering from the atmosphere using a pulsed nitrogen ultraviolet laser. Nature. 1967;**216**:142-143

[15] Melfi SH. Remote measurement of the atmosphere using Raman scattering. Applied Optics. 1972;**11**:1605-1610

[16] Shiina T, Minami E, Ito M, Okamura Y. Optical circulator for in-line type compact lidar. Applied Optics. 2002;**41**:3900-3905

[17] Shiina T, Yoshida K, Ito M, Okamura Y. In-line type micro pulse lidar with annular beam—Experiment. Applied Optics. 2005;**44**:7407-7413

[18] Shiina T, Yoshida K, Ito M, Okamura Y. In-line type micro pulse lidar with annular beam—Theoretical approach. Applied Optics. 2005;**44**:7467-7473

[19] Grishin M, editor. Advances in Solid-State Lasers: Development and Applications, Chapter 8. INTECH; 2010

[20] Shiina T, Oguchi K, Fukuchi T. Polarization-independent optical circulator for high accuracy Faraday depolarization lidar. Applied Optics. 2012;**51**:898-904

[21] Fukuchi T, Shiina T, editors. Industrial Applications of Laser Remote Sensing. Bentham Science Publishers Ltd; 2012

Chapter 4

Investigating the Nature of Insulator-Metal Transition in Neutron-Transmutation-Doped Ge:Ga

Samy Abd-elhakim Elsayed

Additional information is available at the end of the chapter

http://dx.doi.org/10.5772/intechopen.72241

Abstract

n-type germanium samples irradiated with fast reactor neutrons with a fluency range from 2×10^{16} up to 1×10^{20} cm^{-2}. As a result of irradiation, n-Ge samples are converted into p-type Ge. The dc conductivity is measured in wide temperature range from 1.5 up to 300 K. Insulator metal transition occurs at irradiation fluency 5×10^{18} cm^{-2}. The Bohr's radius of localization (a_H) is obtained and found to be equal to 43 Å. Mott variable range hopping and Shklovskii-Efros percolation models are applicable in the present data.

Keywords: germanium, irradiation disordered, insulator-metal transition, Mott variable range hopping, percolation theory of conduction, Anderson transition

1. Introduction

1.1. Ionizing and non-ionizing radiations

Sun light consists of electromagnetic (EM) waves, which provide light and heat. Sunlight electromagnetic waves consist of infrared (IR), visible, and ultraviolet (UV) frequencies; as the color goes more into blue, the wavelength decreases and the frequency increases and vice versa, as the color goes into red color, the wave length increases and frequency decreases. Generally, the EM radiations are divided into two categories depending on wavelength ionizing and non-ionizing. This division depends on the radiation energy or frequency; high frequency radiation (>3×10^{15} Hz) is ionizing radiation and low frequency radiation (<3×10^{15} Hz) is nonionizing radiation [1]. When a substance absorbs the EM radiation with a frequency lower than 3×10^{15} Hz, excitation occurs, and the radiation excites the motion of molecules, or excites an electron from occupied orbital into an empty higher energy orbital. But if the absorbed EM radiation has frequency greater than

3×10^{15} Hz, ionization occurs, that is, removing an electron from an atom or molecule. The atom consists of nucleus and extranuclear structure. The extranuclear structure consists of electrons revolving around the nucleus in certain energy levels; each level contains a restricted number of electrons when an electron jumps from high energy level down to lower energy level; an EM radiation with energy equal to the difference between the two energy levels emits from the atom. When an electron jumps from high energy level to the low energy level, a high frequency $\cong 3 \times 10^{17}$ Hz emits from the atom, this is X-ray radiation. The X-ray radiation is an ionizing radiation. The nucleus consists of protons which have positive charge equal to the electron charge and mass 1.67×10^{-27} Kg. The number of protons in the nucleus equals the number of electrons outside the nucleus and the number of protons or electrons of an atom is denoted by Z (the atomic number) and neutrons which have no electric charge, but have magnetic moment. The number of neutrons is denoted by N. The total number of protons plus neutrons in a nucleus is denoted by A = Z + N (the mass number) [2]. When the nucleus is stable, no irradiation emits from it, but some heavy nuclei excited "not stable" and need to emit some energy to reduce their energies and become more stable [3]. The nucleus emits very high EM radiations, which are gamma rays. Gamma rays have frequency 3×10^{20} Hz, and are ionizing radiations. The unstable nuclei may also emit β particles, protons or neutrons, and/or α-particles, which is the nucleus of He^{++} atom; sometimes, the nucleus emits high energy electromagnetic radiation gamma waves (γ). The stable nucleus is characterized by the mass number A, and the atomic number of protons' number (Z) which equals the number of neutrons (N). In large nuclei $N \cong 1.7\,Z$, repulsion between protons causes instability in the nucleus. This instability decreases if the number of neutrons excesses the number of protons. Instable nucleus, such as U^{238}, emits radiation. This is called natural radioactivity. Nuclear radioactivity involves three kinds of radiations: α, β, and γ radiations. Each of these radiations has distinct properties. Examples of artificial radioactivity are: when α particle collide with aluminum a proton is rejected leaving a stable silicon isotope $Al_{13}^{27} + He_2^4 \rightarrow Si_{14}^{30} + H_1^1$ or a neutron rejected leaving an isotope of phosphorus $Al_{13}^{27} + He_2^4 \rightarrow P_{15}^{30} + n_0^1$ and this phosphorus element is instable and disintegrates into silicon by emission of positron $P_{15}^{30} \rightarrow S_{14}^{30} + e^+$. This type of nuclear reaction has technological application [4], opens the way of peaceful using of nuclear power, in doping semiconductor by nuclear irradiation. Semiconductor industry is playing a very important role for mankind's everyday life and economic growth. The most famous radioactive semiconductor materials' doping is the neutron transmutation doping (NTD). This new technology is used in doping semiconductor materials such as germanium and silicon; the advantages of this new method are (i) homogenous distribution of doping atoms in the host material, (ii) controlling the concentration of doping atoms by controlling the irradiation dose, and (iii) controlling the compensation K in the host material. This new technology opens the way into tailoring materials with definite conductivity at definite temperature for a certain technology application. For example:

$$^{74}Ge(n,\gamma)^{75}Ge \xrightarrow[82\text{min}]{\beta} {}^{75}As + \nu$$

$$^{76}Ge(n,\gamma)^{77}Ge \xrightarrow[11.3hr]{\beta} {}^{77}As + \nu$$

$$^{77}As \xrightarrow[38.3hr]{\beta} {}^{77}Se + \nu$$

$$^{70}Ge(n,\gamma)^{71}Ge \xrightarrow[11.4day]{\varepsilon} {}^{71}Ga + \nu \tag{1}$$

But (NTD) has disadvantages like: (i) irradiation cost, (ii) reduction in minority carrier life time, and (iii) radioactive safeguards considerations.

n-germanium irradiated reactor neutrons and/or γ converted into p-type germanium [5].

2. Effect of ionizing radiation on the electrical properties of semiconductors

The electrical conductivity of solids ranges from $10^9(\Omega.\text{cm})^{-1}$ for pure metal such as copper at liquid helium temperature to at most 10^{-22} $(\Omega.\text{cm})^{-1}$ for the best insulators or nonmetals at the same temperature [6]. These vast differences in electrical conductivity separates the metallic and non-metallic states of matter, this variation amounts to a factor at least 10^{31}, which is probably widest variation for any physical property. Therefore, from electrical conduction characteristics, it is well-known that solid materials are divided into four categories:

i. Metallic conduction (good electrical conductors with electrical conductivity that reaches to $10^9(\Omega.\text{cm})^{-1}$ at liquid helium temperature)

ii. Insulators (Poor electrical conduction with electrical conductivity that reaches to $10^{-22}(\Omega.\text{cm})^{-1}$ at liquid helium temperature)

iii. Semiconductors, this type of materials distinguishes from metals and insulators by their negative coefficient of resistance (their resistance decreases with increasing temperature, while in metals it increases)

iv. Super-conductors, in this class of materials, the resistance vanishes at low temperatures, the temperature at which resistance vanishes depends on the material's type, composition, structure, and preparation method.

This chapter focused on the transition between metallic and insulator phases of conduction. Mott is first one who recognized the problem of metal-insulator transition [7]. He noticed that, although the band theory successfully explained the electronic structure of solids, it explains why some materials have metallic conductivity, other semiconductors, or insulators. But, the band theory gives incorrect ground states of some transition metal oxides such as CoO, MnO, and NiO. In the case of CoO, for example, CoO has NaCl-like structure with a unit cell containing one Co atom and one O atom. The outer shell of the Co atom has an electron configuration of 3d4s, and the O atom 2S2P, that means, the number of electrons per unit cell is 9 + 6 = 15, an odd number. Band theory predicted that crystals with unit cell containing an odd number of electrons should have metallic conductivity behavior. But experimentally it is proved that CoO is actually an insulator with a large gap. These transition metal oxides are called Mott insulators [8] and were first inadequate of band theory. Mott said there is something wrong in band theory. What is that? Mott discussed the band theory in that although it had succeeded in explaining the cause of some materials having metallic conductivity, other semiconductors, and insulators, the band theory neglected the correlation between electrons. These transition metal oxides contain strongly correlated electrons, so the band theory fails in such systems with strongly correlated electrons. Mott started from the model suggested by

Bloch which is called "collective electron model." In this model, the electrons in the outermost shell of the atoms concerned are not supposed to be bound to their individual atoms, but are assumed to be free to move through the lattice under the effect of the atomic cores and the average field of the other electrons. This field is periodic with the period of the lattice. Mott criticized this model that the electrons tend to keep away from one another because of the Coulomb's repulsive force. In metals, each electron carries with it a surrounding sphere, into which other electrons are unlikely to come. In insulators, one of the electrons in outermost shell of an electron will not be able to move to the next atom unless an electron from the next atom gets out to create a free place to it. Mott declares that the collective electron model cannot explain the optical absorption in insulators [8].

For many years Mott insulators were neglected by most authors of solid state physics text books. By the discovery of high T_C superconductors in cuprates and the rediscovery of colossal magnetoresistance in manganite's the situation changed. These materials have proved to be doped-Mott insulators. So, the physics of Mott insulators and doped Mott insulators become important in contemporary research in condensed matter physics. The main problem for Mott insulators is to find the correlation energy of the electrons. It is Hubbard energy U for double occupancy of the same orbital on the same atomic site by two electrons. Mott starts to solve the problem of strongly correlated electrons. He supposed that for an array of electrons centers with cubic structure, there are two possibilities. The first is that the crystal is metal and the second is an antiferromagnetic insulator. But an antiferromagnetic lattice can split the conduction band into full and empty bands, these are called Hubbard bands, the insulating behavior does not depend on antiferromagnetic order, and continues above Neel temperature [9].

3. Mott metal-insulator transition (MIT)

3.1. Transition metals and their alloys

Transition metals are the metals with incomplete d state in atomic structures like Co, Fe, Ni, Cd, etc. In magnetic transition metal compounds, the conduction electrons lie wholly in a d-band, with some extent of hybridization with 4s electrons. This means that the d- band is separated from conduction band and contains an integer number of electrons per atom. And a band containing an integer number of electrons must exceed a certain width if it is to be metallic. Based on Gutzweller [10], approximation of variational calculation method of the ground state wave-function for model Hamiltonian with single-binding band and with only intra-atomic Coulomb interaction between the electrons, considering the case in which one electron per atom and, they found that the Gutzweller variational state is always metallic [11]. The experimental results of Craig et al. [12] on the electrical resistivity of Ni near its Curie temperature lead Fisher et al. [13] to criticize the theoretical treatment of resistive anomaly theory suggested by de Gennes and Friedel [14] because, in their treatment they supposed long-range fluctuations of the magnetization near the critical point and this lead to incorrect result, but the short-range order fluctuations which make the dominant contribution to the temperature dependence of resistivity. Monecke et al. [15] investigated the specific heat of the

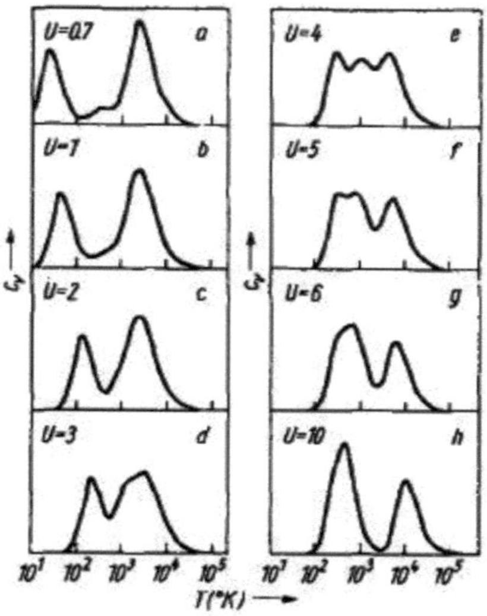

Figure 1. C_v dependence on temperature at different values of Hubbard energy.

Hubbard system which consists of four atoms and four electrons, in wide temperature range from 10 K up to 10^5 K. And, he found that in the specific heat C_v dependence on temperature, as seen in **Figure 1**, there are three maxima, two of them related to Néel transition and sudden change in magnitude of magnetic moment and the third maxima proved as Mott transition [7].

3.2. Mott metal insulator transition

Mott started to explain the metal insulator transition on the basis of Hubbard band model [16]. Mott outlines the properties of a crystalline array of one-electron centers, each described by a wavefunction φ(r) behaving as exp.($-r/a_H$) for large r, at distance a from each other, sufficiently large for the tight binding approximation to be useful. If the number of electron per atom deviates from integral value, the model should always predict metallic behavior, but with one or any integral number of electrons per atom this is not so, and the system is insulating [7]. The most convenient description is in terms of Hubbard intra-atomic energy U which is defined by

$$U = \iint \left(\frac{e^2}{kr_{12}} \left|\phi(r_{12}\right|^2 \left|\phi(r_2\right|^2 d^3x_1 d^3x_2 \tag{2}$$

For hydrogen-like atom

$$U = \frac{5e^2}{8ka_H}$$

If P is the ionization potential of each atom and E is the electron affinity, by assuming that the functions φ are not changed by addition of an extra electron

$$U = P–E \tag{3}$$

The properties of such system are the following:

i. when the distance a between the centers is large, the overlap energy integral ***I*** is small, the system is expected to be antiferromagnetic, with energy below that of the ferromagnetic state equal to $-2zI^2/U = -B/(2zU)$, where B is the band width $B = 2z\mathbf{I}$. The Néel temperature T_N will be such that kT_N is of this order.

ii. An extra electron placed on one of the atoms is able to move with a k vector just as in normal band. It proposed to have an energy in the "upper Hubbard band." It well polarizes the spins on surrounding atoms antiparallel to itself, or parallel if the atomic orbitals are degenerate, forming a "spin polaron." Its bandwidth is probably not very different from B calculated without correction.

iii. A hole formed by taking an electron away from one atom has similar properties, being able to move with wavevector k and having a range of energies not very different from B.

iv. The two bands will overlap when a is small enough, and consequently B great enough, to ensure that

$$B \geq U$$

A metal-insulator transition then occurs, sometimes known as "Mott transition."

The Mott transition is similar to the band crossing transition, being from antiferromagnetic insulator to antiferromagnetic metal. As the overlap increases, the number n of free carriers increases, the moment on the atoms and Neel temperature will tend to zero and disappear when $n = 1/2z$. **Figure 2** shows magnetic moment and Néel temperature at the transition between antiferromagnetism and normal metal against B/U [7], here we have two situations; first, the antiferromagnetism is absent, and after the magnetic moment have disappeared the electron gas becomes highly correlated, this means that small proportion ζ of the sites are doubly occupied at any moment, or unoccupied and the spins of the electrons on other sites

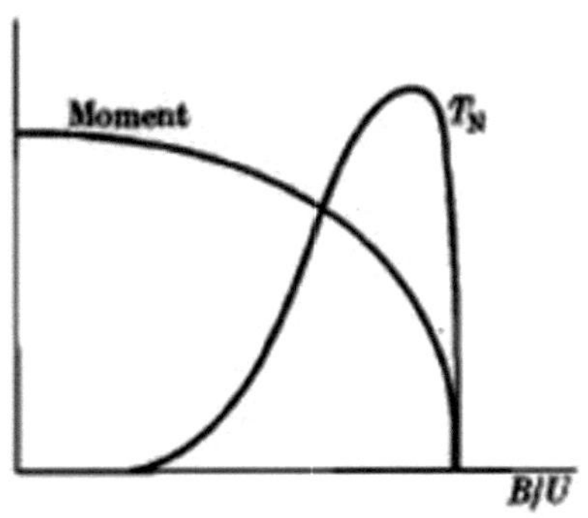

Figure 2. Magnetic moment and Nèel temperature at transition antiferromagnetism and normal metal as function of B/U.

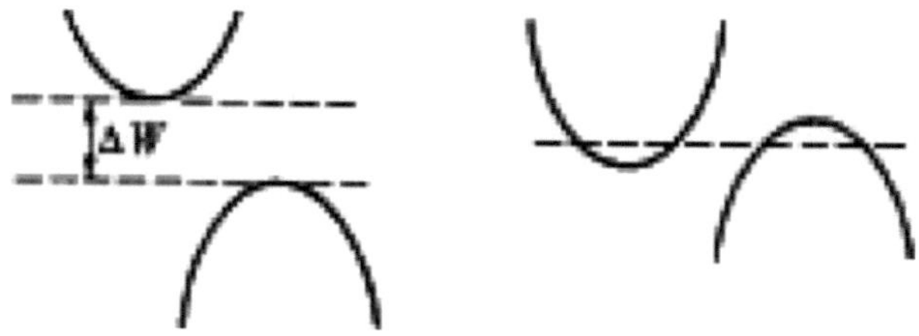

Figure 3. Two Hubbard bands overlapped in metal case, and not overlapped in insulator non-metal.

are no longer arranged antiferromagnetically, but resonate quantum mechanically between their two positions. The band is no longer split into two Hubbard bands **Figure 3**; instead there is a large enhancement by $1/2\zeta$ of the effective mass. The simplest form of metal-insulator transition is the form of semi-metal, a material for which the conduction and valence band overlap, to a non-metal, where there is no overlap.

In the antiferromagnetic insulator, the magnitude of the Hubbard gap is not greatly affected when the temperature goes through the Neel point T_N. The Hubbard gap is then, essentially, the energy required to take an electron from an atom and put it on a distant atom where another electron is already present. The energy U needed to achieve this does not depend on coexistence of the antiferromagnetic long-range order. At zero temperature, an insulator can be described in terms of the full and empty bands determined by long-range antiferromagnetic order. In simple cubic lattice of one electron centers, if $U/B > 0.27$, a sharp metal-insulator transition occurs above Néel temperature. But it has been found that above Néel temperature, for a rigid lattice, the transition between an antiferromagnetic insulator and a metal as U/B varies is not sharp; Mott thought that this is a result of considering the long-range forces in Hubbard model [16]. As the temperature increases [17], the electrons are excited into the conduction band, their coupling with moments lowers the Néel temperature. Thus, the disordering of the spins with consequent increase of entropy is accelerated. Ramirez et al. [18] showed that a first-order transition to degenerate gas in the conduction band, together with disordering of the moments, is possible.

At the transition, the value of $B \geq U$ can be evaluated for hydrogen like wavefunction, $U = \frac{e^2}{8ka_H}$ since both U and B are known

$$I = I_o e^{-\alpha R}$$
$$I_o = \left\{\frac{3}{2}(1 + \alpha R) + \frac{1}{6}(\alpha R)^2\right\} e^2 \alpha \qquad (4)$$

I is the overlap energy integral

$$\alpha = \frac{(2mW_0)^{1/2}}{\hbar}$$

and α is defined as the distance at which the single-well wavefunction $\exp.(-\alpha R)$ falls off. Since the value of B near the transition is given by

$$B = 23z\left(\frac{e^2\alpha}{k}\right)e^{-\alpha R} \tag{5}$$

but because of the rapid variation of $e^{-\alpha R}$ and αR depend on z; $\alpha R = 5.8$ [28], and if z the coordination number = 6

$$n^{1/3}a_H \cong 0.2 \tag{6}$$

This equation is used for the discussion of metal insulator transition in doped semiconductors.

3.3. Anderson insulator metal transition

Anderson [19] showed that at a certain randomness, the electron wavefunction becomes localized. According to the Anderson model, as shown in **Figure 4**, there exists a critical value V_O/B above which the diffusion is impossible at zero temperature. Anderson introduced a random potential into each well within limits $\pm\frac{1}{2}V_o$. This leads to finite mean free path l such that $a/l = 0.7(V_O/B)^2$. If $V_O/B \cong 1$, the value of l approaches unity. Mott and Kavah [17] showed that it cannot be smaller, and the wavefunction has the form

$$\Psi_{ext} = \sum_n C_n \exp(i\varphi_n)\psi_n \tag{7}$$

where C_n are coefficients, φ_n are random phases, as the V_O/B increases greater than unity the phases still random, the scatter of the coefficients C_n will increase, and the wave function becomes localized having the form

$$\Psi = Rral\left[\psi_{ext}\right]\left\{e^{\frac{-(r+r_0)}{\zeta}}\right\} \tag{8}$$

where r_o is some point in space, and ζ is the localization length. The critical value of V_o/B depends on the co-ordination number z. Mott and Kavah [17] supposed that V_o/B greater than unity, the range of energies in the band increased to $(v_o^2 + B^2)^{0.5}$, so they write in the mid-gab

$$N(E) = 1/\left\{a^3\left(v_o^2 + B^2\right)^{0.5}\right\} \tag{9}$$

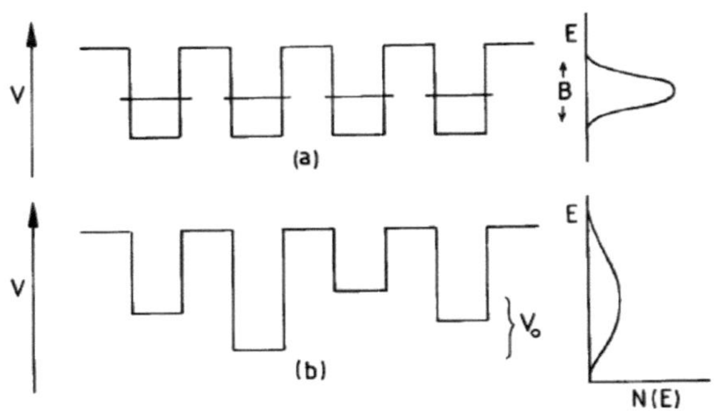

Figure 4. The potential energy of an electron in the Anderson model: (a) without a random potential, and (b) with a random potential V_O. B is the band-width in case (a). The density of states N(E) is shown.

3.4. Mott versus Anderson explanation of metal-insulator transition

Mott treated the metal-insulator transition problem by considering the electron gas as screening the positive charge on the donors, so the potential energy seen by electrons in heavily doped semiconductors is of the form

$$V(r) = \frac{e^2}{kr} e^{-qr} \tag{10}$$

where k is the dielectric constant of the semiconductors, and q is the screening constant depending on the electron density n. Mott also proposed that for an electron which its potential energy follows formula v(r) = (e^2/kr) exp.(−qr) and if q is large enough the electron gas would have metallic properties. But as the density of electron gas drops, bound states become possible. By calculation, the value of n at which the transition occurs to be

$$n^{\frac{1}{3}} a_H = 0.25 \tag{11}$$

where a_H is the Bohr radius of donors. It is predicted that a discontinuous transition would occur from a state at which all electrons were bound to donors to one in which they were free, this is Mott transition. Mott [7] predicted that at the transition, the indirect band gap should disappear. This proved in germanium and silicon. In these semiconductor materials, in case of electrons and holes production by irradiation, electrons and holes combine at low temperature to form excitons, which are for indirect band gap have comparatively long-life time. Excitons attract each other and droplets of an electron-hole gas formed. The electron-hole gas droplets have two forms; one is the excitonic insulator, or an electron-hole metallic gas. Mott and Kavah [17] argued that if V_o/B is less than critical value, the states are localized up to mobility edge. As shown in **Figures 4** and **5**, the localization length ξ tends to infinity as E → E_c, if we write a/ξ = const. $(E_c - E)^s$, neither the constant nor the index is known with certainly.

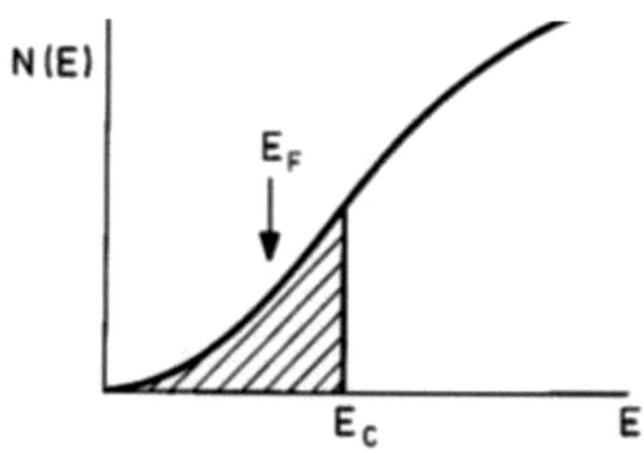

Figure 5. Density of states with potential energy of **Figure 4** showing the mobility edge E_c. The position of the Fermi energy E_F below E_C is shown.

3.5. Conductivity of a metal in Anderson model

The conductivity of a metal is due to a gas non-interacting electron in the field of Anderson random potential. The conductivity of a metal is usually written in the form $\sigma = ne^2\tau/m$, where n is the number of electrons per unit volume, m is the effective mass, and τ is the time of relaxation. If l is the mean free path, S_F is the Fermi surface area $4\pi k_F$, and $n = (8\pi/3)k_F^3/8\pi^3$, σ takes the form $\sigma = S_F e^2 \iota/12\pi^3\hbar$, Mott and Kavah [17] examined the last formula mean free path l = a, at z = 6 n = $1/a^3$, hence $K_F a = K_F l = (3\pi^2)^{1/3} \cong 3.1$, the conductivity is thus:

$$\sigma \cong \frac{1}{3}\left(\frac{e^2}{\hbar a}\right) \tag{12}$$

This last equation assumes a spherical Fermi surface; if a = 2 Å, the numerical value is 3000 $(\Omega.\text{cm})^{-1}$. But Mott and Kavah expected the smallest conductivity in the range 2–5000 $(\Omega.\text{cm})^{-1}$.

As the ratio V_o/B increases the conductivity decreases because

i. As the density of states at E_F decreases, the conductivity decreases, and may be written in the form $\sigma = \frac{1}{3}\left(\frac{e^2}{\hbar a}\right)g^2$,

ii. Where,

$$g = \frac{B}{\left\{1.75\left(B^2 - V_o^2\right)^{\frac{1}{2}}\right\}}$$

In this case a minimum metallic conductivity expected, and given by

$$\sigma_{\min} = \frac{e^2}{3\hbar a} g_c^2 \approx 0.03 \frac{e^2}{\hbar a,}, \quad g_c \cong 1/3 \tag{13}$$

g_c is the critical value at Anderson localization

iii. As the ratio V_o/B increases, another phenomenon appears which is called incipient localization, this phenomenon decreases the conductivity, $\sigma \to 0$ as T tends continuously to zero, as V_o/B increases to the critical value.

The Anderson transition is a phenomenon in which some systems for example, doped semiconductors, the zero temperature Fermi energy move from point above the mobility edge to a point below it. Generally, σ goes continuously to zero, as shown in **Figure 6a**, but if the transition is induced by magnetic field H, very sharp transition occurs as shown in **Figure 6b**. The conductivity at zero-field mobility edge E_c (0) is given by

$$\begin{aligned} &\sigma(E_C(0)) = 0.03\frac{e^2}{\hbar L}, L = L_H = \left(\frac{c\hbar}{eH}\right)^{\frac{1}{2}} L > a \\ &L = a \ldots\ldots\ldots\ldots\ldots\ldots\ldots\ldots\ldots L_H < a \end{aligned} \tag{14}$$

As a conclusion of this section, in Mott model, the metal insulator transition occurs because the correlated electron gas cause increasing in the distance between the atomic centers in the lattice. There is a minimum metallic conductivity given by

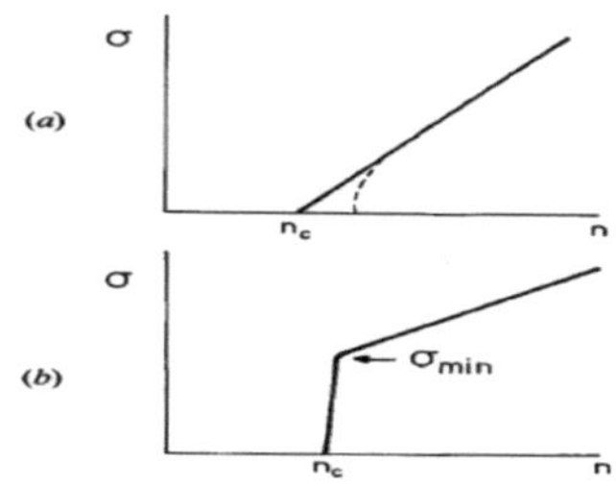

Figure 6. The behavior of conductivity at zero temperature of a degenerate electron gas as a function of charge carrier concentration n, so the E_F moves through the mobility edge E_C, (a) if σ_{min} does not exist and (b) if it does. The broken line shows the possible effects of interactions.

$$\sigma_{\min} = \frac{0.03e^2}{\hbar L_i}$$

where L_i is the inelastic diffusion length, and the transition is a first-order transition. Meanwhile, Anderson transition metal insulator occurs because the existence of disorder, as the disorder increasing the wave function becomes nonlocalized (metallic state) and the transition is continuously occurring, and becomes sharp only under the effect of external magnetic field, and there is no minimum metallic conductivity.

4. Experimental part

Samples of germanium doped with Arsenic irradiated with fast neutrons with energies of $E \geq 0.1$ MeV in the range of $2 \times 10^{16} \leq \Phi \leq 1 \times 10^{20}$ cm^{-2}. As a result of irradiation, all the original samples become disordered p-type [20–23]. In order to reduce the transmutation doping effect, all samples were placed in 1-mm-thick cadmium containers. During irradiation in the reactor, the ratio between thermal neutrons fluency and the fluency of fast neutrons was about 10. So, it was possible to obtain samples of germanium "doped" with acceptor-like radiation defects {Ge (RD)}. To ensure that the electrical properties were controlled by the transmutation doping, a complete annealing at 450°C for 24 h was performed. For electrical resistivity measurements, a special double wall glassy cryostat is designed [24]. This cryostat is attached with vacuum pump its evacuation rate is faster than the evaporation rate of He^4 gas, thus the pressure inside the cryostat is decreased and hence the temperature. The conventional four probe method is used for electrical resistivity measurements. Ni electrode is participated in the desired position on the samples using electrochemical deposition technique (cold method). Thin Cu wires are fixed above the Ni electrodes using In. The samples were in parallelepiped shape with length about 8–12 mm, thickness about 1–2 mm, width about 2–3 mm, and the electrode apart about 3–4 mm resistivity of Ge (RD) was measured in the temperature range from 1.7 up to 300 K. The temperature was determined with a semiconductor thermistor in the interval 77.4–4.2 K, from saturated vapor pressure of He^4 in the interval 4.2–1.5 K. The voltage across samples always less than 1 V and the current across the sample decreases from μA to nA order as the temperature lowered. The electrical properties of Ge (RD) are determined solely by acceptor-like radiation defects [25, 26]. The least square fitting method, using a computer program Excel 2010, is used to analysis the data extracted and to calculate the uncertainty in the obtained results.

5. Experimental results

Figure 7 shows the resistivity dependence on temperature of germanium samples irradiated with different doses of neutron irradiation. From this figure, the conductivity activation energy of conductivity ($\sigma = \rho^{-1}$) is obtained, from the empirical equation argued by Fritzche [27]

$$\rho = \rho_o \exp. - (E_C - E_F)/KT + \rho_1 \exp. - (E_A - E_F + W_1)/KT + \rho_2 \exp.(W)/KT \quad (15)$$

where the term (ρ_o exp.-(E_C-E_F)/KT) represents transport by carriers excited beyond the mobility edge into non-localized (extended) states at E_C or E_V. And ρ is the resistivity, ρ_o is the pre-exponential factor. A plot of lnρ versus 1/T will yield a straight line if E_C-E_F is a linear function of T over the temperature range measured. As the temperature decreases, transport carriers excited into localized states at the band edges and the resistivity is given by ρ_1 exp.-(E_A-E_F + W_1)/KT, where W_1 is the activation energy for carriers hopping, W_1 should decrease with decreasing temperature [28] On the account of variable-range nature of the hopping transport. However, as the temperature dependence is through the carrier activation term, approximately linear dependence of ln ρ versus 1/T is again expected. As the temperature lowered more, there

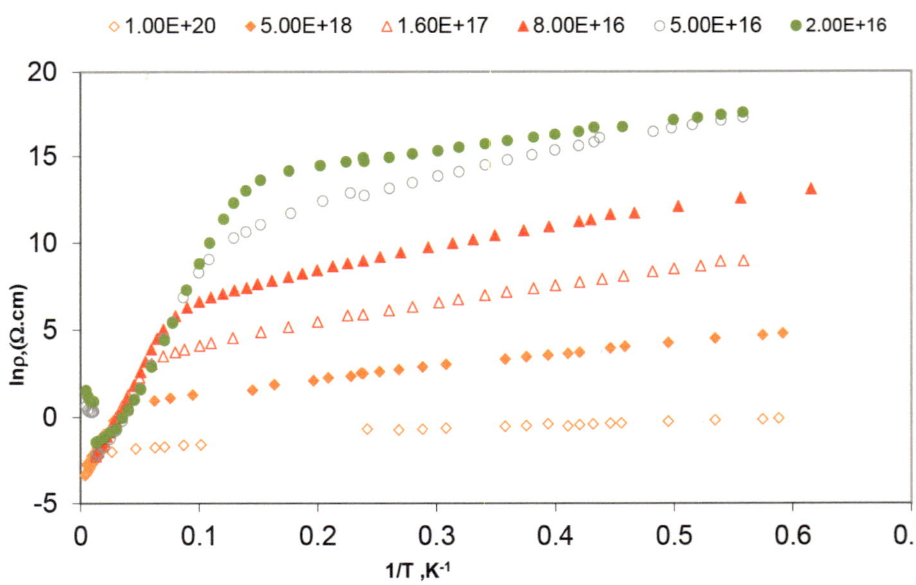

Figure 7. ln σ = f (1/T).

will be a contribution from carriers with energies near E_F which can hop between localized states. This contribution is described by ρ_2 exp. (-W)/KT, where $\rho_1 \leq \rho_2$ and W is the hopping energy of the order of half the width of the band of the states. The extracted results from **Figure 7** are summarized in **Table 1**.

F is the neutron flux, P_{300}, P_{77} are the number of charge carriers at 300 K and 77 K respectively where E_1, E_2, E_3, are the activation energies of different conductivity mechanisms. And m_{300}, m_{77}, are the mobility of charge carriers at 300 K and, 77 K respectively.

The obtained results reveal that $E_1 > E_2 > E_3$, and $\rho_o<\rho_1<\rho_2$ for all samples.

Figure 8 shows the dependence of resistivity on square root of temperature. This figure is required to apply Shklovskii's percolation theory of conduction where [29]

$$\rho = \rho_o e^{\left(\frac{T_{ES}}{T}\right)^{05}} \tag{16}$$

From this equation, T_{ES} is obtained for each irradiation dose

$$T_{ES} = \frac{\beta e^2}{ka}$$

where β is a constant, e is the electronic charge, k is the dielectric constant, and a is the localization radius.

Figure 9 shows the dependence of resistivity on $T^{-0.25}$. From this figure [30], Mott variable range hopping is applied to obtain Mott characteristic temperature from the equation

Φ, cm^{-2}	p300	logp300	p77	E_1, meV	E_2, meV	E_3, meV	Logμ_{300}	Logμ_{77}
0	1.6E + 15	15.20	1.5E + 15	21.2	----	----	3.59	4.32
1E + 15	1.5E + 15	15.18	3.3E + 14	28.9	----	----	3.42	4.24
5E + 15	2.5E + 15	15.4	9E + 14	22.6	----	2.27	3.45	4.24
2E + 16	5E + 15	15.7	2.23E + 15	17.8	0.73	0.79	3.39	4.17
5E + 16	1.15E + 16	16.06	6.56E + 15	11.5	1.93	0.84	3.4	3.87
8E + 16	3.7E + 16	16.57	1.18E + 16	10.7	1.57	0.92	3.45	3.64
1.6E + 17	7.4E + 16	16.87	2.2E + 16	9.2	0	0.9	3.45	3.53
5E + 18	1.4E + 17	17.14	4.05E + 16	6.4	----	0.63	3.48	3.23
1E + 20	4.18E + 17	17.62	1.6E + 17	3.7	0	0.03	2.99	2.69

Table 1. The concentration of holes at 300 K and at 77 K, the conductivity activation energies of the three stages of conductivity, E_1 which is the band to band transition activation energy, E_2 which is activation energy of inter band transition activation energy, E_3 which is the hopping conduction activation energy, the mobility μ of holes at 300 K and 77 K, respectively.

$$\rho = \rho_o e^{\left(\frac{T_M}{T}\right)^{0.25}} \tag{17}$$

where $T_M = \frac{\beta_o}{N(E_F)a^3}$, where β_o is a constant, $N(E_F)$ is the density of states at Fermi level, and a is Bohr radius of localization.

The obtained values of T_M and T_{ES} from **Figures 8** and **9** are given in **Table 2**.

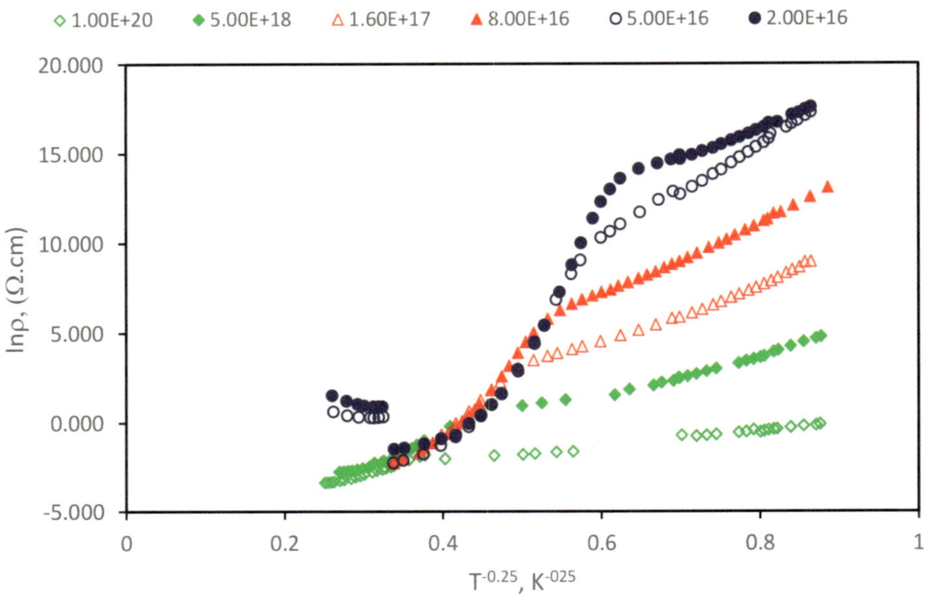

Figure 8. ln σ = f (1/$T^{0.5}$).

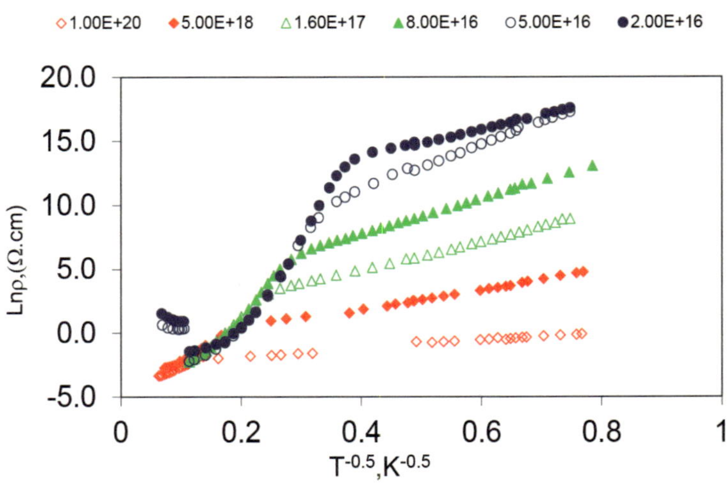

Figure 9. ln (σ) = f(1/$T^{0.25}$).

Φ, cm^{-2}	T_M, K	T_{ES}, K
1.2×10^{17}	148,454	151.50
8×10^{16}	201,906	194.94
5×10^{16}	262,951	320.4
2×10^{16}	69,232	113.89

Table 2. Mott characteristic temperature and Efros-Shklovskii characteristic temperature at each irradiation dose.

Table 2 indicates that T_M is much higher than T_{ES}, this table also clears the coexistence of Mott variable range hopping mechanisms of conduction, and Shklovskii and Efros percolation mechanism of conduction, the coexistence of both models explained the low temperature conduction mechanisms reported [31, 32] in different compositions. But it is not clear up to now, are these two mechanisms of conduction simultaneously exist or consecutively one mechanism dominates in a certain temperature range and then, the other one. **Figure 10** shows the dependence of σ_o on the irradiation dose ϕ, the experimentally obtained data are fitted with Eq. (17)

$$\ln \phi = \frac{\ln \sigma_o}{a + b\ln \sigma_o + c\sqrt{\ln \sigma_o}} \tag{18}$$

where a, b, and c are coefficients to fit present experimental data, a = 777.8, b = 18, and c = 238.4. **Figure 10** shows that the pre-exponential factor ρ decreases with increase the irradiation fluency until reaching to minimum value and start to increasing this behavior could be explained as follows, as the irradiation dose increases the charge carrier increases, which causes a diminution of the resistivity until the concentration of charge carrier reaches its critical

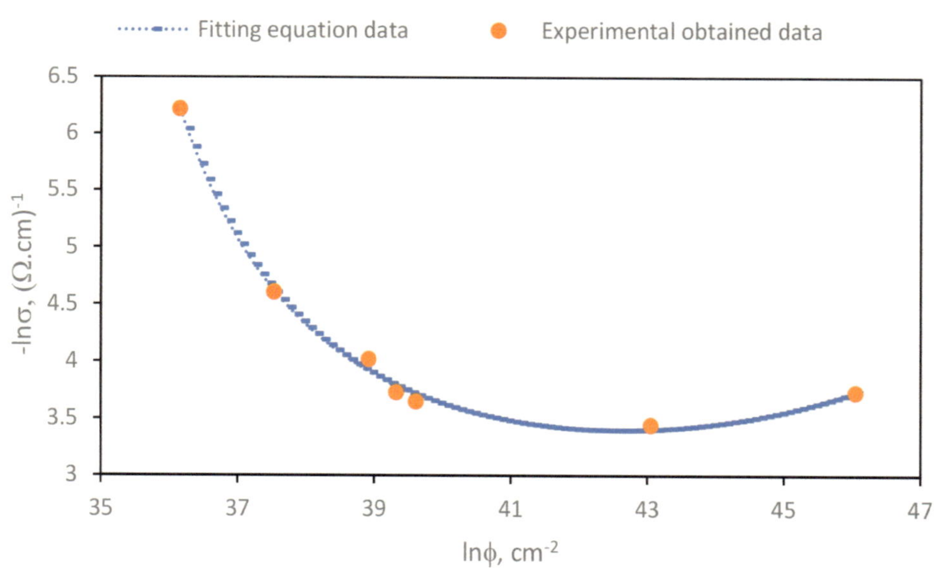

Figure 10. Fitting of Eq. (17) for the dependence of pre-exponential factor σ_o on the irradiation flux (φ).

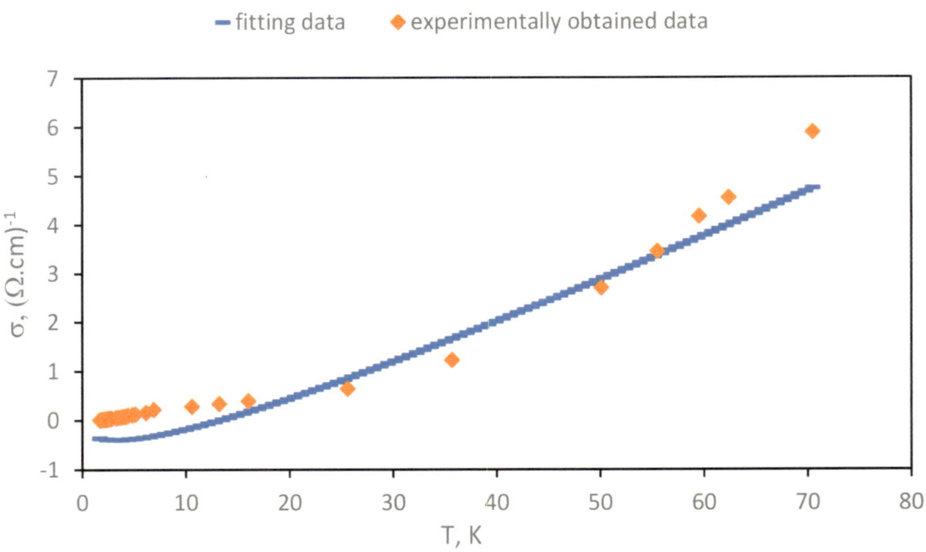

Figure 11. Fitting of Eq. (3) for dependence of conductivity on temperature of sample irradiated 5×10^{18} cm^{-2}.

value at which sample behaves as metallic like conductivity. In metallic conductors, the charge carrier augmentation enhances the possibility of collision between the charge carriers and with the doping centers, this increases the resistivity.

Figure 11 shows the fitting of conductivity dependence on temperature of sample irradiated with dose of 5×10^{18} cm^{-2} to equation [33]

$$\sigma = \sigma(0) + aT^{0.5} + bT \tag{19}$$

where σ(0) is the residual conductivity at 0 K, the second and third term are related to weak localization effect. The fitting parameters as obtained from this equation are σ(0) = 0, a = −0.37, and b = 0.12.

Figure 12 shows the fitting of conductivity dependence on temperature of sample irradiated with irradiation dose 1×10^{20} cm^{-2} using Eq. (19). The fitting parameters as obtained from this equation are σ(0) =0, a = 0.65, and b = 0.065.

Since we have insulator metal transition, this means that the sample starts as an insulator, but because of irradiation the same samples becomes a metallic-like conductor after certain irradiation dose. This means that the conduction in a material may behave like semiconductor (insulator) or metallic characteristic of conduction. The same equation, Eq. (19), did not fit the conductivity temperature dependence of sample irradiated with dose 1.6×10^{17} cm^{-2}. This means that this sample is in the insulator side of conductivity. However, the sample irradiated with dose 5×10^{18} is in the metallic conductivity side. For sample with irradiation dose 5×10^{18} cm^{-2}, the fitting parameter value is −0.037, the negative value of parameter a is reported in [34]. The negative value of a in some samples appears only at the beginning of metallic conductivity. From experimental data analysis of **Figure 7**, the extrapolation intercept with ordinate axes gives the resistivity, σ_o value. For sample irradiated with dose 1×10^{20} and

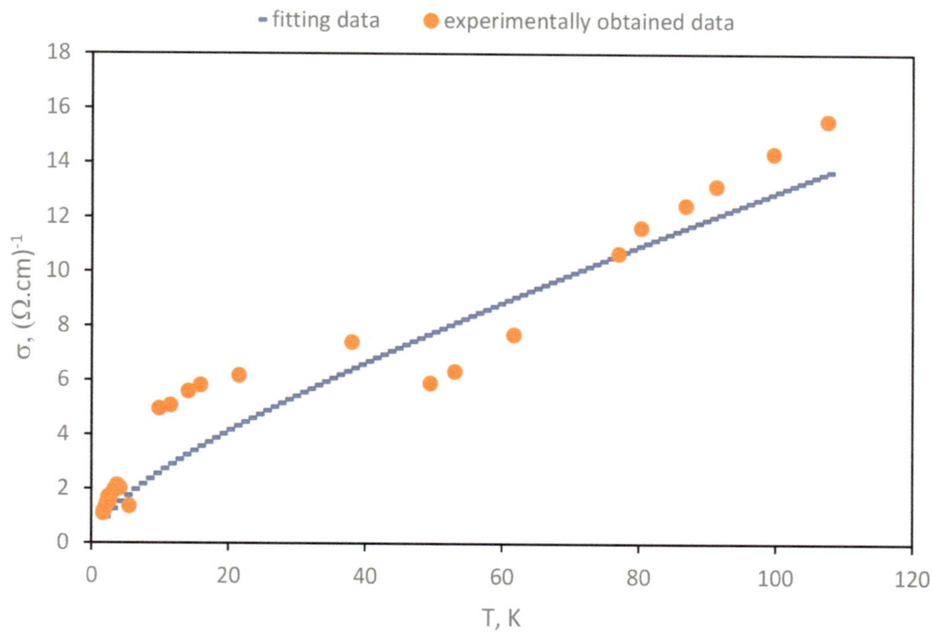

Figure 12. Fitting of dependence of conductivity on temperature of sample irradiated 1×10^{20} cm^{-2}.

1.6×10^{17} cm^{-2}; $\sigma_o = 0.024$ and $\sigma_o = 0.032$ $(\Omega\text{-cm})^{-1}$, from these values, one can say that σ_o in the metallic side of conductivity is dependent on irradiation dose. **Figure 10** shows the dependence of pre-exponential factor on neutron irradiation dose, it is clear that as the neutron irradiation dose increases, the value of the pre-exponential factor decreases. It reaches its minimum value at neutron irradiation dose 1.6×10^{17} cm^{-2} and starts to increase. Mott predicted that the minimum metallic conductivity $\sigma_{min} = \frac{0.026e^2}{\hbar a}$. In case of a = 2 Å, he predicted minimum metallic conductivity to be of the range 2–5000 $(\Omega.cm)^{-1}$, and he considered that this is the smallest conductivities to be expected, unless the density of states broadened by disorder. In this work samples irradiated with neutron flux 1×10^{20} and 5×10^{18} cm^{-2}, having minimum metallic conductivity equals to 0.024 and 0.032 $(\Omega\text{-cm})^{-1}$, respectively, these values are much smaller than the minimum value of metallic conductivity predicted by Mott. This may be due to the high disorder in germanium caused by neutron irradiation, or the Mott minimum metallic conductivity satisfied in materials just began its metallic conductivity behavior.

Figure 13 shows the dependence of $\ln(\rho_3)$ on average impurity center distance as predicted by Mott variable range hopping model, the relation between ρ_3 on the average impurity center distance is

$$\rho_3 = \rho_{03} e^{\frac{\alpha}{N_A a}}, \ \alpha = 1.74 \tag{20}$$

where ρ is the resistivity, α is a constant, N_A is the concentration of acceptor impurity centers per cm^{-3}, and a is the Bohr localization radius. From the line slope of **Figure 13**, the Bohr radius of localization is calculated and found equal to 43 Å, which is very close to the value of Bohr radius of localization at insulator metal transition which is 40 Å.

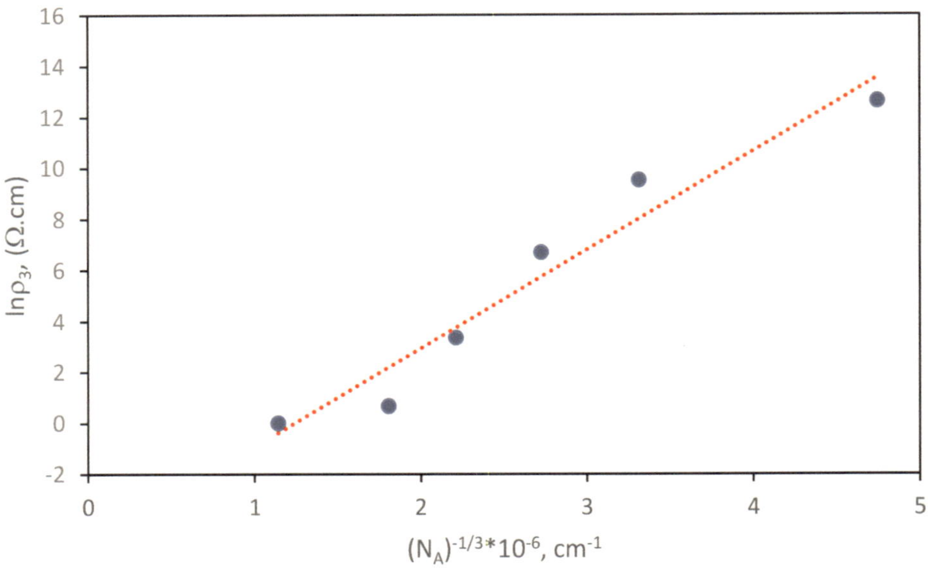

Figure 13. The dependence of ln $\rho_3$on average impurity center distance $(N_A)^{1/3}$.

6. Conclusions

The overall obtained results reveal that the n-type germanium is converted into p-type germanium by a fast reactor neutron irradiation. From analysis of the dependence, the electrical conductivity on temperature, the activation energies of conduction at different conductivity mechanism that; $E_1 > E_2 > E_3$ and the pre-exponential factors $\rho_2 > \rho_1 > \rho_0$ were observed for each irradiation dose. But these values decrease with increasing irradiation dose. The metallic-like behavior starts at irradiation dose 5×10^{18} cm^{-2}. There are sweeping between Mott variable range hopping model and Efros-Shklovskii percolation model. Each model dominates in certain range of temperature. From analysis of Mott, variable range hopping the Bohr radius of localization at Insulator metal transition is obtained to be equal to 43 Å, and the obtained minimum metallic conductivity value is 0.024 $(\Omega\text{-cm})^{-1}$, which is much lower than the value predicted by Mott theory of insulator-metal transition.

Author details

Samy Abd-elhakim Elsayed

Address all correspondence to: samy.abdelhameed@science.bsu.edu.eg

New Materials and Renewable Energy Laboratory, Physics Department, Faculty of Science, Beni-Suef University, Beni-Suef, Egypt

References

[1] Hecht E. Optics. San Francisco, Boston, New York: Adison Wesley; 2002

[2] Das A, Ferbal T. Nuclear and Particle Physics. Toh Tuck Link, Singapore: World Scientific Publishing; 2003

[3] Haas EW, Martin JA. Nuclear transmutation doping from the view point of radioactivity formation. In: Messe JM, editor. Neutron Transmutation Doping in Semiconductors. New York and London: Plenum Press; 1979

[4] Heller EE. Isotopically Controlled Semiconductors. 2004. http://escholarship.org/uc/item/1621k38s

[5] Zabrodskii AG, Alekseenko MV. Fermi level scan spectroscopy of gap states in Ge and Si–Ge alloysbased on the kinetics of neutron transmutation doping. Physica B. 2006;**376-377**:253

[6] Edwards PP, Johnston RL, Rao CNR, Tunstall DP, Hensel F. The metal-insulator transition: A perspective. Philosophical Transactions of the Royal Society of London A. 1998;**356**

[7] Mott NF. Metal Insulator Transition. 2nd ed. London: Taylor & Francis; 1990, and references therein

[8] Duan F, Guojin J. Introduction to Condensed Matter Physics. Vol. VI. Toh Tuck Link, Singapore: World Scientific Publishing Company; 2005

[9] Mott NF. The electrical conductivity of transition metals. Proceedings of the Royal Society of London A. 1936;**153**:699

[10] Gutzwiller MC. Effect of correlation on the ferromagnetism of transition metals. Physics Review. 1964;**134**:4A

[11] Brinkman WF, Rice TM. Application of gutzwiller's variational method to the metal-insulator transition. Physical Review B. 1970;**2**:10

[12] Craig PP, Goldburg WJ, Kitchens TA, Budnick JI. Transport properties at critical points: The resistivity of s. Physical Review Letters. 1967;**19**:23, and references therein

[13] Fisher ME, Langer JS. Resistive anomalies at magnetic critical points. Physics Review. 1968;**20**:13

[14] deGennes PE, Friedel J. Anomalies de résistivité dans certains métaux magníques. Journal of Physics and Chemistry of Solids. 1958;**4**

[15] Heing KH, Monecke J. Conductivity of the four-point Hubbard Model. Physica Status Solidi. 1972;**49**:K139

[16] Hubbard J. Theory the connexion with many-body perturbation electron correlations in narrow energy bands. VI. Proceedings of the Royal Society of London A. 1967;**296**:100

[17] Mott NF, Kaveh M. Metal Insulator transitions in non-crystalline systems. Advances in Physics. 1985;**34**:3

[18] Ramirez R, Fallicov LM, Kimball J. Metal-insulator transitions: A simple theoretical model. Physical Review B. 1970;**2**:8

[19] Anderson PW. Absence of diffusion in certain random lattices. Physics Review. 1958;**109**: 1492

[20] Crawford JH Jr, Cleland JW. Nature of bombardment damage and energy levels in semiconductors. Journal of Applied Physics. 1959;**30**(6):1204

[21] Crawford JH Jr, Lark-Horovitz K. Fast neutron bombardment effects in germanium. Physics Review. 1950;**78**:815

[22] Cleland JW, Crawford JH, Pigg JC. Transmutation-produced germanium semiconductors. Physics Review. 1950;**78**:814

[23] James HM, Lark-Horovitz K. Localized electronic states in bombarded semiconductors. Zeitschrift für Physikalische Chemie. 1951;**198**:107

[24] Fritzche H, Lark Horovitz K. The electrical properties of germanium semiconductors at low temperature. Physica A. 1954:834-844

[25] Dobrego VP, Ermolaev OP, Tkachev VD. Investigation of shallow radiation defect levels produced in Ge by fast neutron irradiation. Physica Status Solidi. 1977;**44**:435

[26] Konopleva RF, Novikov SR. Measuring the relative fast-neutron flux distribution in the VVR-M reactor with semiconductor detecting elements. Atomnaya Energiya. 1961;**11**:545. in Russian

[27] Fritzche H. Resistivity and hall coefficient of antimony doped germanium at low temperature. Journal of Physics and Chemistry of Solids. 1958;**6**:69

[28] Mott NF, Davis EA. Electronic Processes in Non-Crystalline Materials. Oxford: Clarendon Press; 2012

[29] Shklovskii BI, Efros AL. Electronic Properties of doped Semiconductors. Berlin, Heidelberg, New York, Tokyo: Springer-Verlag; 1984

[30] Elhakim SA, Morsy MA. Simultaneous $\gamma(Co^{60})$-quanta irradiation and isothermal annealing-induced insulator–metal transition in a-$As_4Se_4Te_2$ chalcogenide composition. Radiation Effects and Defects in Solids. 2014;**169**:4

[31] El-Sayed A. Fractal explanation of Meyer-Neldel rule. Journal of Non-Crystalline Solids. 2017;**458**:137-114

[32] El-hakim SA. Observation compensation effect and crossover between Percolation and variable-range-hopping regimes for dc conductivity in germanium irradiated with large fluencies fast neutrons. Journal of Optoelectronics and Advanced Materials. 2014;**16**(11–12): 1367

[33] Kawade K, Suzuki A, Tanka K. Electronic structure and electrical conductivity of amorphous Si–Ti alloys manifesting the metal-insulator transition. Journal of the Physical Society of Japan. 2000;**69**(3):777

[34] Hussey NE, Tanaka K, Takagi H. Universality of the Mott–Ioffe–Regel limit in metals. Philosophical Magazine. 2004;**84**(24):2847

Chapter 5

Radiation-Induced Degradation of Organic Compounds and Radiation Technologies for Purification of Aqueous Systems

Igor E. Makarov and Alexander V. Ponomarev

Additional information is available at the end of the chapter

http://dx.doi.org/10.5772/intechopen.72074

Abstract

Environmental application of radiation technologies is an important part of radiation processing. Radiation treatment of aqueous systems contaminated with organic compounds is a promising method of water and wastewater purification and corresponding technologies are being developed. In this chapter, the following aspects of radiation treatment process are considered: sources of contamination and major contaminants of water and wastewater; primary processes in aqueous systems initiated by ionizing radiation; principal ways of contaminant conversion as consequences of primary processes (complete mineralization of organic compounds, partial decomposition of organic molecules resulted in detoxification, decolorization, disinfection of polluted water, and improvement in biological degradation of contaminant, polymerization of monomers' contaminants, oxidation-reduction processes, and coagulation of colloids); sources of ionizing radiation; and main equipment applied in radiation technologies of aqueous system purification.

Keywords: radiation chemistry, radiation technology, wastewater, electron accelerators, organic compounds

1. Introduction

Changes in a matter resulted from reformation of chemical bonds structure under the action of ionizing radiation gave rise to implementation of various radiation technologies. At present, an overwhelming majority of the latter represent modification of materials, that is, polymers curing, grafting, production of thermo-shrinkable plastics, improvement of semiconductor circuits, and so on [1, 2]. At the same time, there exists an important area of radiation technologies application, namely environment conservation, where those technologies are not widespread as yet, or just are at the stage of development. One of the problems in environment conservation is purification of flue gases and wastewater, and radiation technologies possess a high potential in the direction providing for higher purification degree, unachievable in some cases by conventional purification technologies.

Traditionally, radiation treatment (by gamma/X-rays or accelerated electrons beams) for the purification of water and wastewater is attributed to the so-called Advanced Oxidation Processes (AOPs) and corresponding technologies—to Advanced Oxidation Technologies (AOTs) (see, e.g., books [3, 4]). The reason is that initially new processes, appeared in addition to routine purification processes, such as biological treatment, coagulation, sedimentation, and filtration, have been based on the oxidation of organic impurities by strong oxidizing agent—hydroxyl radicals, produced firstly in chemical reactions with participation of hydrogen peroxide, then with the use of physical methods, and irradiation of aqueous systems produces hydroxyl radicals as one of the main intermediates. However, the attribution of radiation treatment to AOPs seems to be not quite correct, since ionizing radiation initiates in irradiated media not only oxidizing processes but also reducing ones. Though the formation of reducing particles under radiolysis, as well as under photolysis, sonolysis, electrolysis, and so on, has been mentioned earlier repeatedly [3–5], the role of reductive processes in the purification of water and wastewater was not considered as specific and important one, partly due to comparatively low importance of the processes in the decomposition of organic compounds.

A significant advantage of radiation-induced processes over other AOPs consists in sufficient knowledge of mechanism of the processes including available qualitative and quantitative data on primary events of energy absorption, intermediates formation ways and properties, and reactions of organic compounds conversions. Nowadays, new research and review articles on radiation treatment are being published which demonstrate high interest to the problem of environmental application of radiation technologies.

The aim of this chapter is to show interrelation between well-established primary radiation-chemical processes and the main ways of organic contaminants conversions, as well as to present the current state of radiation equipment being used at present or proposed to be used in radiation technologies of water and wastewater purification.

2. Types of aqueous systems to be purified

The application of radiation technologies to the purification of aqueous systems to a certain extent is determined by a subject of radiation treatment, that is, composition of the system. There are several types of aqueous systems to be purified commercially which differ considerably in qualitative and quantitative composition. The latter is usually expressed either chemically as a concentration of components (in units of moles or mass of the component per volume or mass of solution) or in units characteristic of water/wastewater purification technologies, that is, COD (Chemical Oxygen Demand), BOD (Biological Oxygen Demand), and TOC (Total Organic Carbon).

TOC is determined as the total mass of carbon atoms in dissolved organic compounds per unit solution volume and expressed usually in mg/L (mg/dm^3) or in ppm (parts per million, i.e., mass of carbon per mass of solution), the latter being numerically almost the same, as far as a mass of 1 dm^3 of diluted aqueous solution is close to 1 kg. COD is a mass of oxygen necessary for complete oxidation of organic compounds, present in unit volume of the solution, into carbon dioxide (and other inorganic compounds), while the BOD is a mass of oxygen, consumed in the process of biological oxidation of organic compounds, present in unit volume of the solution. Since biological oxidation is a long-term process, instead of BOD the notation BOD_5 is often used, which means BOD measured after 5 days of biological oxidation. Like TOC, both COD and BOD are expressed in the same units and do not relate to the kind of organic compounds polluting water or wastewater, but rather reflect the total organic matter content and its part that can be degraded in biological processes [6].

The main types of aqueous systems to be purified are considered to be water (from different sources) and wastewater.

2.1. Water

Water, consumed in the course of human activity, besides municipal water, that is, water of potable quality offered by water companies, has two natural sources: (1) groundwater, that is, water, seeped through from the surface and present in porous rocks below the surface, shallow wells, or deep aquifers, and (2) surface water, that is, water from a source that is exposed to the environment like rivers, canals, lakes, and open wells [7].

Contamination of water may occur due to following reasons: municipal water—from pipelines, fittings, biofouling; groundwater—due to failing of septic systems, leaking sewer lines, and from land discharge by passage through soils and fissures or interaction with surface water; surface water—by treated wastewater, discharge of raw sewage, municipal wastewater, storm-water runoff, runoff from urban and agricultural areas, also animals and humans are both indirect and direct contributors to the contamination.

The level of water contamination usually is not high to use it for many purposes. Thus, most pristine rivers will have a 5-day carbonaceous BOD below 1 mg/L. Moderately polluted rivers may have a BOD value in the range of 2–8 mg/L. Rivers may be considered severely polluted when BOD values exceed 8 mg/L [8].

Groundwater from depth and confined aquifers is usually microbially safe and chemically stable in the absence of direct contamination; however, shallow or unconfined aquifers can be subject to contamination from discharges or seepages associated with agricultural practices (e.g., pathogens, nitrates, and pesticides), on-site sanitation and sewerage (pathogens and nitrates), and industrial wastes.

Both groundwater and surface water have natural impurities including inorganic salts and, particularly—surface water, NOM (natural organic matter), represented mainly by humic and fulvic acids, cellulose, lignin) which have to be removed, or reduced in concentration, when using the water for drinking purposes [9, 10]. Quality specifications for drinking water are subject of policies, regulations, and standards on both national and international levels [11–14]. Guidelines are being established that determine the limit concentration of chemical compounds contaminating potable water. Now, there are more than 100 compounds for which guideline values have been established (see **Table 1**, data compiled from [11]).

Guideline values for the chemicals listed in **Table 1** are in the range of 10^{-1}–10^{-4} mg/L [15], and real contamination of water exceeds these values only in extraordinary cases when special purification is needed.

Much more significant problem of water purification is microbial contamination [11, 15]. **Table 2** shows the ranges of microbial contamination of surface and groundwater.

Exact guideline values for microbial contamination are not established, in [15] being just noted that "*E. coli* or thermotolerant coliform bacteria must not be detectable in any 100 ml sample." However, report [16] quotes data that for tolerable drinking water densities of Enteric virus, *Giardia*, and *Cryptosporidium* are equal to 2.2×10^{-7}, 6.8×10^{-6}, and 1.7×10^{-6} per liter, correspondingly. It means that to make water from natural sources drinkable, the concentration of microorganisms should be reduced by more than six orders of magnitude.

So, using water from natural sources for industrial purposes does not require its purification, but using it as potable water requires disinfection and in special cases removal of hazardous contaminants.

2.2. Wastewater

Contamination level of wastewater is much higher than that of water and the goal of wastewater purification, except for the problems of reuse [13], is to reduce contamination down to

Chemicals from industrial sources, human dwellings	Chemicals from agricultural activities	Chemicals used in water treatment or materials in contact with drinking water
Inorganic		
Cd, Hg, Cyanide	Nitrate/nitrite	Sb, Cu, Pb, Ni
Organic		
Benzene; Carbontetrachloride; Di(2-ethylhexyl)phthalate; 1,2-Dichlorobenzene; 1,4-Dichlorobenzene; 1,2-Dichloroethane; 1,1-Dichloroethene; 1,2-Dichloroethene; Dichloromethane; 1,4-Dioxane; Edetic acid; Ethylbenzene; Hexachlorobutadiene; Nitrilotriacetic acid; Pentachlorophenol; Styrene; Tetrachloroethene; Toluene; Trichloroethene; Xylenes	Alachlor; Aldicarb; Aldrin and dieldrin; Atrazine; Carbofuran; Chlordane; Chlorotoluron; Cyanazine; 1,2-Dibromo-3-chloropropane; 1,2-Dibromoethane; 1,2-Dichloropropane; 1,3-Dichloropropene; Dichlorprop; Dimethoate; Endrin; Fenoprop; Isoproturon; Lindane; MCPA; Mecoprop; Methoxychlor; Metolachlor; Molinate; Pendimethalin; Simazine; Terbuthylazine; Trifluralin	**Disinfectants:** Monochloramine; Chlorine **Disinfection byproducts:** Bromate; Bromodichloromethane; Bromoform; Chloral hydrate; Chlorate; Chlorite; Chloroform; Cyanogen chloride; Dibromoacetonitrile; Dibromochloromethane; Dichloroacetate; Dichloroacetonitrile; Formaldehyde; Monochloroacetate; Trichloroacetate; 2,4,6-Trichlorophenol; Trihalomethanes **Pesticides:** Chlorpyrifos; DDT (and metabolites); Pyriproxyfen **Organic contaminants from treatment chemicals:** Acrylamide; Epichlorohydrin **Organic contaminants from pipes and fittings:** Benzopyrene; Vinyl chloride

Table 1. Contaminants in water for which guideline values have been established (naturally occurring chemicals: As, Ba, B, Cr, F, Mn, Mo, Se, U).

level appropriate for discharge into environment. Usually, wastewater is divided into two main categories: industrial and municipal or domestic, in accordance with its specific composition. However, amenably to more strict definition adduced in [13]: "Wastewater is liquid waste discharged from homes and other residential premises, commercial and industrial premises and similar sources, to individual disposal systems or to municipal sewer pipes. It contains mainly human excreta and used water. Wastewater collected in municipal sewerage systems is called municipal wastewater or municipal sewage." A similar definition is contained in [17]: "Municipal wastewater originates from domestic, industrial, commercial and institutional sources within a given human settlement or community. Urban wastewater includes both municipal wastewater and urban runoff." It means that industrial wastewater may be regarded as a part of municipal one. Nevertheless, while considering wastewater composition, more useful is to break down wastewater by sources. **Table 3** lists the data on contamination of wastewater from different sources [17].

From the data of **Table 3**, it follows that the basic constituent of municipal and domestic wastewater is organic matter, including nutrients, which is substantially biodegradable. Special purification is required to remove pathogenic microorganisms and emerging pollutants. On the contrary, the composition of industrial wastewater differs markedly for different industries (see **Table 4**), the level of organic contamination is high, and organic matter is often almost non-biodegradable.

Pathogen or indicator group	Lakes and reservoirs	Impacted rivers and streams	Wilderness rivers and streams	Groundwater
Campylobacter	20–500	90–2500	0–1100	0–10
Salmonella	—	3–1000	1–4	—
E. coli (generic)	10,000–1,000,000	30,000–1,000,000	6000–30,000	0–1000
Viruses	1–10	30–60	0–3	0–2
Cryptosporidium	4–290	2–480	2–240	0–1
Giardia	2–30	1–470	1–2	0–1

Table 2. Examples of high detectable concentrations (per liter) of enteric pathogens and fecal indicators in different types of source waters.

Sources of wastewater	Typical components
Domestic wastewater	Human excreta (pathogenic microorganisms), nutrients, and organic matter. May also contain emerging pollutants (e.g., pharmaceuticals, drugs, and endocrine disruptors)
Municipal wastewater	Very wide range of contaminants, such as pathogenic microorganisms, nutrients and organic matter, heavy metals, and emerging pollutants
Industrial wastewater	Contaminants depend on the kind of industry

Table 3. Typical components of wastewater from different sources.

Industry type	Wastewater generation, m^3/ton	Typical content of effluent	COD, g/L
Alcohol refining	16–32	High level of BOD and COD	5–22
Beer and malt	5–9	BOD, COD, SS, nitrogen, phosphorus—variable by individual processes; pH variable due to acid and alkaline cleaning agents; high temperature	2–7
Coffee	No data	High levels of BOD and SS concentrations;	3–15
Dairy products	3–10	Dissolved sugars, proteins, fats, and additive residues; BOD, COD, SS, nitrogen, and phosphorus	1.5–5.2
Fish processing	8–18	High levels of BOD and SS concentrations; strong organics, antibiotics, growth hormones, pesticides, and insecticides	~2.5
Meat and poultry	8–18	High levels of BOD and SS concentrations; strong organics, antibiotics, growth hormones, pesticides, and insecticides	2–7
Organic chemicals	0–400	Pesticides, pharmaceuticals, paints and dyes, petro-chemicals, detergents, plastics, feed-stock materials, byproducts, product material in soluble or particulate form, washing and cleaning agents, solvents, and added-value products such as plasticizers	0.8–5
Petroleum refineries	0.3–1.2	Production of fossil fuels—contamination from oil and gas wells and fracking; hot cooling water	0.4–1.6
Plastic and resins	0.3–1.2		0.8–5
Pulp and paper	85–240	Chlorinated lignosulfonic acids, chlorinated resin acids, chlorinated phenols and chlorinated hydrocarbons—about 500 different chlorinated organic compounds identified; colored compounds and absorbable organic halogens (AOX); pollutants characterized by BOD, COD, suspended solids (SS), toxicity, and color	1–15
Starch production	10–1.5–42	High levels of BOD and SS concentrations;	
Sugar refining	4–18	High levels of BOD and SS concentrations;	1–6
Vegetable oils	1–5	High levels of BOD and SS concentrations; high particulates, some dissolved organics, surfactants;	0.5–1.2
Vegetables and fruits	7–35	High levels of BOD and SS concentrations; high particulates, some dissolved organics, surfactants;	2–10
Wine and vinegar	11–46	High levels of BOD and SS concentrations; high particulates, some dissolved organics, surfactants	0.7–3.0

Table 4. Wastewater characteristics in some industries (compiled from data of [17]).

The content of organic matter in wastewater from different sources varies, reaching on average up to several grams per liter. Thus, quoted in [18] the mean values for wastewater in France and USA are, respectively, the following: COD—300–1000 and 250–1000, BOD—100–400 and 110–400, suspended solids—150–500 and 100–350, total potash and nitrogen—30−100 and 20–85, and total phosphorus—1–25 and 4–15. More than twofold difference between COD and BOD shows that organic matter cannot be removed by biological treatment and additional treatment of wastewater is required.

3. Principal ways of radiation-induced chemical processes promoting aqueous system purification

Impurities in water and wastewater are present, as a rule, at relatively very low concentration, except for some kinds of wastewater with very high content of organic compounds. Electron fraction of inorganic and organic substances in ground and surface water, as well as in industrial and municipal/domestic wastewater, usually does not exceed 1%. Since the absorption of ionizing radiation, both electromagnetic one and accelerated electrons in the range of energies 0.5–10 MeV, by a particular component of a mixture, is proportional to the electron fraction of the component, and almost all the radiations are absorbed by water resulting in primary processes of water radiolysis, the products of which (mainly—hydrated electrons e^{-}_{aq}, hydrogen atoms $^{\bullet}H$, and hydroxyl radicals $^{\bullet}OH$) do react with substances dissolved or dispersed in water. As a whole, the process of water purification with the use of ionizing radiation is presented in **Figure 1**.

In this section, the main processes induced by ionizing radiation, which can be applied to water purification technologies, are discussed together with a brief discussion on the formation and properties of the products of water radiolysis. Most principal reactions of reactive products of water radiolysis with a solute, concerning wastewater treatment by ionizing radiation, have been overviewed recently [5].

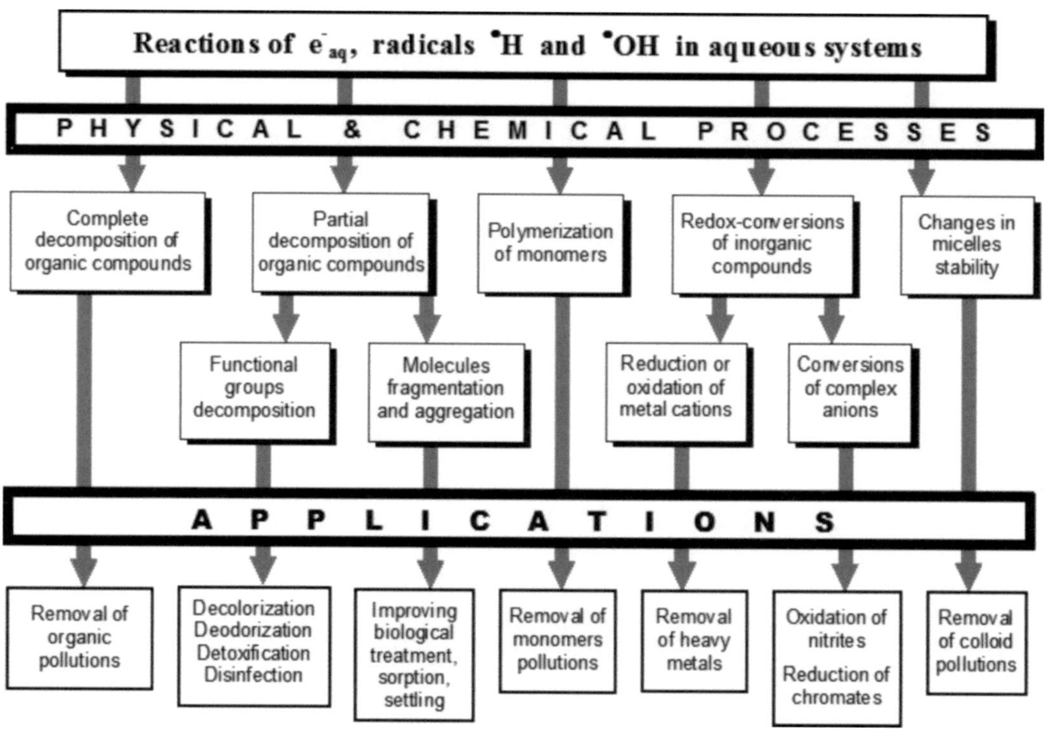

Figure 1. Main processes of radiation purification of aqueous systems.

3.1. Initial stages of ionizing radiation action on aqueous systems

As it has been mentioned earlier, reactive products of water radiolysis are the initiators of almost all the processes of dissolved or dispersed substance conversion in such aqueous systems as water and wastewater. The mechanism of primary processes in radiolysis of water including ionization and excitation of molecules by ionizing radiation, formation, and subsequent reactions of the products of radiolysis is well established both qualitatively and quantitatively [19–21]. Briefly, it can be described as the following:

3.1.1. Energy absorption

Ionizing radiation, such as electromagnetic one (gamma or X-rays) and accelerated electrons in the range of energies 0.5–10 MeV, is absorbed in the liquid medium by portions of near 30 eV in small volumes (referred to as "spurs") separated by about 100 nm. Because of high local concentration of reacting species formed in "spurs," their reactions with each other are weakly influenced by a solute.

3.1.2. Primary processes

Primary processes of ionizing radiation interaction with water molecules include

Excitation and ionization of water molecules, decay of singlet excited states, and primary particles formation ($\leq 10^{-14}$ s).

$$H_2O \longrightarrow H_2O^*$$

$$H_2O \longrightarrow e^- + H_2O^+$$

$$H_2O^* \longrightarrow {}^\bullet H + {}^\bullet OH$$

$$H_2O^* \longrightarrow H_2 + O$$

Formation of hydrated electrons e^-_{aq}, relaxation of ionized molecules of water ($\leq 10^{-12}$ s)

$$e^- \longrightarrow e^-_{therm} \longrightarrow e^-_{aq}$$

$$H_2O^+ + H_2O \longrightarrow H_3O^+ + {}^\bullet OH$$

Main reactions of formed particles in "spurs" ($\leq 10^{-7}$ s)

$$e^-_{aq} + H_3O^+ \longrightarrow {}^\bullet H + H_2O$$

$${}^\bullet H + {}^\bullet H \longrightarrow H_2$$

$${}^\bullet H + {}^\bullet OH \longrightarrow H_2O$$

$${}^\bullet OH + {}^\bullet OH \longrightarrow H_2O_2$$

$$e^-_{aq} + {}^\bullet OH \longrightarrow OH^-$$

Product	Reactive				No reactive			
	e^-_{aq}	$^\bullet$H	$^\bullet$OH	O	H_2	H_2O_2	H_3O^+	OH^-
G, μmol/J	0.28	0.06	0.28	<0.01	0.04	0.07	0.33	0.05

Table 5. G-values of water radiolysis products at the end of "spur" processes.

3.1.3. *"Initial" yields*

After diffusive expansion of "spurs," when the distribution of formed particles becomes nearly uniform, further reactions proceed in the bulk of solution according to the so-called homogeneous kinetics, the concentrations of water radiolysis products being determined by their "initial" radiation-chemical yield—G-value, that is, number of moles of a product formed per 1 joule of absorbed energy of ionizing radiation (see **Table 5**).

3.1.4. *Influence of oxygen*

In the presence of dissolved oxygen, the following fast reactions take place:

$$e^-_{aq} + O_2 \longrightarrow {}^\bullet O_2^- \qquad k = 1.9 \times 10^{10}\ \mathrm{L\ mol^{-1}\ s^{-1}}$$

$$^\bullet H + O_2 \longrightarrow HO_2^\bullet \qquad k = \sim 2 \times 10^{10}\ \mathrm{L\ mol^{-1}\ s^{-1}}$$

the acid-base equilibriums being established:

$$^\bullet OH \rightleftarrows {}^\bullet O^- + H^+ \qquad pK_A = 11.9$$

$$HO_2^\bullet \rightleftarrows {}^\bullet O_2^- + H^+ \qquad pK_A = 4.7$$

$$H_2O_2 \rightleftarrows HO_2^- + H^+ \qquad pK_A = 11.6$$

$$(^\bullet H \rightleftarrows e^-_{aq} + H^+) \qquad pK_A = 4.4$$

3.1.5. *Reactions in the bulk of solution*

Mutual reactions of water radiolysis products go on in the bulk of solution as homogeneous ones (see **Table 6**), competing thereby with their reactions with a solute.

Properties of reactive free-radical products from water and their reactivity to some specific functional groups in organic compounds are listed in **Tables 7** and **8**.

3.1.6. *Chemical properties of reactive radicals*

Chemical properties of reactive radicals determining their reactions with a solute may be presented as the following:

Hydrated electron as a reducing particle is characterized by the following fast reactions of one-electron transfer from its hydrated state to a molecule/ion in

Reaction	k, L mol^{-1} s^{-1}	Reaction	k, L mol^{-1} s^{-1}
Hydrated electrons		Hydrogen atoms	
$e^-_{aq} + H_2O \rightarrow {}^{\bullet}H + OH^-$	19	${}^{\bullet}H + H_2O \rightarrow {}^{\bullet}OH + H_2$	10
$e^-_{aq} + H_3O+ \rightarrow {}^{\bullet}H + H_2O$	2.3×10^{10}	${}^{\bullet}H + H_2O_2 \rightarrow {}^{\bullet}OH + H_2O$	9×10^7
$e^-_{aq} + H_2O_2 \rightarrow {}^{\bullet}OH + OH^-$	1.1×10^{10}	${}^{\bullet}H + OH^- \rightarrow e^-_{aq} + H_2O$	2.2×10^7
$e^-_{aq} + HO_2^- \rightarrow {}^{\bullet}O^- + OH^-$	3.5×10^9	${}^{\bullet}H + {}^{\bullet}H \rightarrow H_2$	1.2×10^{10} (2 k)
$e^-_{aq} + {}^{\bullet}H \rightarrow H_2 + OH^-$	2.5×10^{10}	${}^{\bullet}H + {}^{\bullet}OH \rightarrow H_2O$	7×10^9
$e^-_{aq} + e^-_{aq} \rightarrow H_2 + 2\,OH^-$	5×10^9 (2 k)	${}^{\bullet}H + HO_2^{\bullet} \rightarrow H_2O_2$ (2 ${}^{\bullet}$OH)	~1×10^{10}
$e^-_{aq} + {}^{\bullet}OH \rightarrow OH^-$	3.0×10^{10}	Oxide radical ions	
$e^-_{aq} + {}^{\bullet}O^- \rightarrow 2\,OH^-$	2.2×10^{10}	${}^{\bullet}O^- + H_2O \rightarrow {}^{\bullet}OH + OH^-$	1.8×10^6
$e^-_{aq} + HO_2^{\bullet} \rightarrow HO_2^-$	1.5×10^{10}	${}^{\bullet}O^- + H_2 \rightarrow {}^{\bullet}H + OH^-$	8×10^7
$e^-_{aq} + {}^{\bullet}O_2^- \rightarrow 2\,HO_2^-$	1.3×10^{10}	${}^{\bullet}O^- + H_2O_2 \rightarrow {}^{\bullet}O_2^- + H_2O$	~5×10^8
Hydroxyl radicals		${}^{\bullet}O^- + HO_2^- \rightarrow {}^{\bullet}O_2^- + OH^-$	4×10^8
${}^{\bullet}OH + H_2 \rightarrow {}^{\bullet}H + H_2O$	4.2×10^7	${}^{\bullet}O^- + O_2 \rightarrow O_3^-$	3.6×10^9
${}^{\bullet}OH + H_2O_2 \rightarrow HO_2^{\bullet} + H_2O$	2.7×10^7	${}^{\bullet}O^- + {}^{\bullet}O_2^- \rightarrow 2\,OH^- + O_2$	6×10^8
${}^{\bullet}OH + HO_2^- \rightarrow {}^{\bullet}O_2^- + H_2O$	7.5×10^9	Perhydroxyl/superoxide radical anions	
${}^{\bullet}OH + OH^- \rightarrow {}^{\bullet}O^- + H_2O$	1.3×10^{10}	$HO_2^{\bullet} + H_2O_2 \rightarrow {}^{\bullet}OH + H_2O + O_2$	~1
${}^{\bullet}OH + {}^{\bullet}OH \rightarrow H_2O_2$	7×10^9 (2 k)	(${}^{\bullet}O_2^- + H_2O_2 \rightarrow {}^{\bullet}OH + OH^- + O_2$)	~1
${}^{\bullet}OH + {}^{\bullet}O^- \rightarrow HO_2^-$	~2×10^{10}	$2\,HO_2^{\bullet} \rightarrow H_2O_2 + O_2$	7.6×10^5
${}^{\bullet}OH + HO_2^{\bullet} \rightarrow O_2 + H_2O$	6×10^9	$HO_2^{\bullet} + {}^{\bullet}O_2^- \rightarrow HO_2^- + O_2$	8.9×10^7
${}^{\bullet}OH + {}^{\bullet}O_2^- \rightarrow O_2 + OH^-$	8×10^9	$2\,{}^{\bullet}O_2^- \rightarrow HO_2^- + OH^- + O_2$	0.3

Table 6. Homogeneous reactions in the bulk of solution.

	e^-_{aq}	${}^{\bullet}$H	${}^{\bullet}$OH	${}^{\bullet}O^-$
Charge	−1	0	0	−1
Optical absorption band:				
λ_{max}, nm	720	<188	235	240
ε_{max}, m^2/mol	1850	<160	60	~30
$W_{1/2}$, eV	0.85	—	~1.5	1.5–2
Partial molar volume, m^3/mol	7×10^{-6}	4×10^{-6}	6×10^{-6}	8×10^{-6}
Diffusion coefficient D, m^2/s	5×10^{-9}	7×10^{-9}	2×10^{-9}	1×10^{-9}
Standard redox potential E^0, V	−2.87	−2.3	1.9	1.7

Table 7. Properties of radicals from water.

- *Inorganic compounds with*
  - all the metal cations, except for cations of alkaline and alkali-earth metals;
  - majority of oxygen-containing and other complex anions, like NO_2^-, NO_3^-, CrO_4^{2-}, $Cr_2O_7^{2-}$, MnO_4^-, $Fe(CN)_6^{3-}$, and so on (low reactivity to SO_4^{2-}, ClO_4^-, CO_3^{2-}, and some others).

Functional group			e^-_{aq}	$^•H$	$^•OH$	Functional group		e^-_{aq}	$^•H$	$^•OH$
Saturated carbon		>CH_2	– –	+	++	Amine	$-NH_2$	– –	– –	++
Unsaturated carbon in:	Alkenes	>C=C<	– –	++	++	Nitro	$-NO_2$	++	+	– –
	Aromatics	>C=C<	– –	++	++	Hydroxy	>CHOH	– –	+	++
	Azines	>C=N–	++	+	++	Ether	–O–	– –	+	++
Nitrile		–C≡N	+	– –	+	Chloro	–Cl	+	–	– –
Carbonyl		>C=O	++	– –	+	Bromo	–Br	++	+	– –
Carboxylic acids		–C(OH) = O	+	– –	+	Iodo	–I	++	++	+
Ester		–C(OR) = O	+	– –	+	Thiol	–SH	++	+	++
Amide		$-C(NH_2) = O$	+	– –	+	Disulfide	–S–S–	++	+	++

Table 8. Reactivity of radicals from water to some specific functional groups in organic compounds (high ++, medium +, low –, and very low – –).

- *Organic compounds with*
  - benzene ring in aromatic compounds;
  - heteroatomic double or triple bonds (carbonyl, nitro, nitrile);
  - thiol, disulfide, halide, nitro functional groups in both saturated and unsaturated (including aromatic) hydrocarbons.

Hydrogen atom reacts with the solute in the following types of reactions:

- Reduction reactions (one-electron transfer $H^• + M^{n+/-} \rightarrow H^+ + M^{(n-1)+/-}$) with
  - majority of metal cations (however, low reactivity to Cd^{2+}, Zn^{2+}, some rare-earth metal cations, etc.);
  - some strong oxidizing inorganic anions, like CrO_4^{2-}, $Cr_2O_7^{2-}$, MnO_4^-;
  - disulfide, iodo, bromo, nitro functional groups, and benzene ring in organic compounds.
- Addition reactions to a double bond ($^•H + >C{=}C< \rightarrow >CH{-}^•C<$) in unsaturated hydrocarbons, aromatic compounds.
- Reactions of H-atom abstraction from saturated carbon ($^•H + >CH- \rightarrow H_2 + >{}^•C-$) in any organic compounds.

Reactions of the last type are formally the oxidizing ones. Also, H-atom can oxidize metal cations in acid solutions, for example, $H^• + Fe^{2+} (+H^+) \rightarrow H_2 + Fe^{3+}$.

Hydroxyl radical is a strong oxidizer and reacts with the solute in oxidizing reactions of the following types:

- Direct oxidation reactions (one-electron transfer), including those ones with
  - all transition metal cations in lower oxidation state, majority of inorganic anions;
  - benzene ring in aromatic compounds and some functional groups;

- Addition reactions to a double bond ($^{\bullet}OH$ + > C=C< → >C(OH)–$^{\bullet}$C<) in unsaturated hydrocarbons, aromatic compounds;
- Reactions of H-atom abstraction from saturated carbon ($^{\bullet}OH$ + > CH– → H_2O + > $^{\bullet}$C–) in any organic compounds.

Kinds of the reactions of e^{-}_{aq}, $^{\bullet}H$, and $^{\bullet}OH$, listed in subdivision 3.1.6, determine the ways of conversions of compounds present in an aqueous system which are considered subsequently in this section.

3.2. Complete mineralization of organic compounds

In principle, every organic compound in aqueous solution can be transformed into stable inorganic substances, such as CO_2 and H_2O, and, if other elements are present, also into ammonium, nitrogen and sulfur oxides, halogenides, and so on, under sufficient absorbed dose of ionizing radiation, the process being called "mineralization." Because of very high energy demand, this process is hardly applicable in practice to significantly reduce the amount of organic matter in a wastewater. However, it may be successfully used to remove hazardous pollutants from wastewater when their concentrations are relatively low.

Radiation-induced process of any organic compound destruction includes C–C bonds rupture reactions (it is necessary in the case of cyclic/aromatic compounds) and consecutive reactions of terminal carbon atom oxidation into carboxyl group followed by decarboxylation, that is, elimination of the atom in the form of CO_2. A simplified scheme of the process looks like this:

(a) *Rapture of* C–C *bond (opening the ring).*

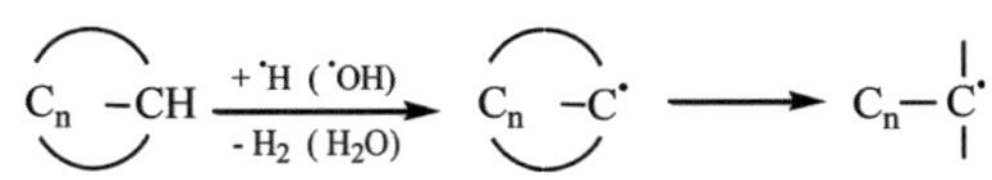

(b) *Oxidation (by* O_2 *attachment) and decarboxylation.*

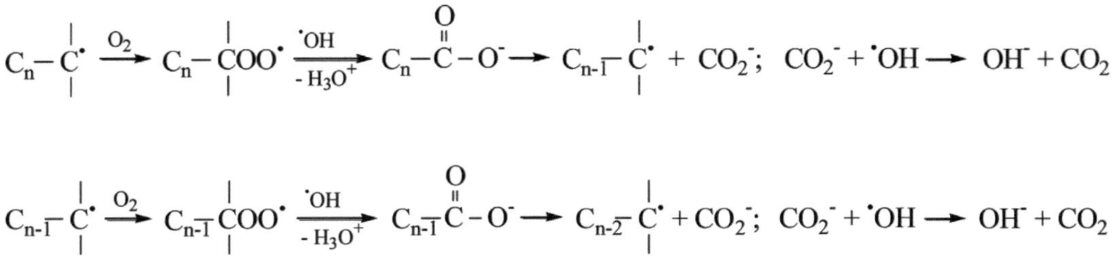

$$C_{n-2} \rightarrow C_{n-3} + CO_2, \text{etc.}$$

In reality, the mechanism of mineralization is much more complicated, especially for aromatic and heteroatomic compounds, but anyway it is a multistage one, and the more composite is organic compound, the more energy it takes to mineralize it completely. Since the process of mineralization proceeds essentially via oxidation reactions, the presence of dissolved oxygen is important because of two main reasons: oxygen converts reducing radicals (e^-_{aq}, $^\bullet H$) into oxidizing ones ($^\bullet O_2^-$, $HO_2^\bullet$) and attaches itself to carbon-centered radicals, providing thereby the formation of carboxyl group. However, in the absence of dissolved oxygen complete decomposition of organic compounds also takes place while the yield of the process is much lower.

The degree of organic compounds mineralization in aqueous solutions upon the action of ionizing radiation can be easily controlled by TOC measurements. As TOC is a mass of all carbon atoms in organic matter, the yield of decarbonization (or mineralization) G_{miner}, expressed in moles of eliminated carbon atoms per 1 kGy of absorbed dose D, is $G_{miner} = (1/12)\times d\mathrm{TOC}/dD$. It makes it possible to compare radiation-chemical yield of carbon elimination with that of initial organic compound degradation G_{degr}, as well as with the yields of radical products from water G_R, being initiators of degradation process. In the absence of radiation-induced chain processes, always the following inequation is right $G_R > G_{degr} > G_{miner}$. Numerous experimental data concerning various organic compounds destruction under irradiation confirm this relationship. Thus, from the data of recent researches it follows that the values of G_{degr} and G_{miner} for irradiation of some organic compounds solutions equals, respectively, for 2-naphthalenesulfonate [22]: 0.14 and 0.08, for monuron [23]: 0.15 and 0.08, for amphetamine [24]: 0.33 and 0.06, for phenol [25]: 0.25 and 0.05, for ibuprofen [26]: 0.25 and 0.05 μmol/J. Low yields of mineralization (0.05–0.08 μmol/J) compared to the yield of OH radicals (0.28 μmol/J) show that in different mechanisms of radiation-induced conversions of different compounds, the elimination of one carbon atom demands three to six radicals $^\bullet OH$, while the conversion of one molecule of initial compound occurs under the action of just one to two radicals. Maximal yields of mineralization, like quoted earlier, are reached at sufficient concentration of dissolved oxygen that requires some time aerating the solution during irradiation [27]. Also, it has been noted repeatedly (see, e.g., [24, 28–31]) that the presence of inorganic anions, such as Cl^-, HCO_3^-, HSO_4^-, reduces the yield of mineralization, mainly due to scavenging the $^\bullet OH$ radicals. A decrease in G_{miner} takes place when subjects of radiation treatment are real aqueous systems (water or wastewater), rather than model solutions prepared with purified water.

In spite of low values of G_{miner}, experimental data obtained (see, e.g., a review [32]) indicate that absorbed doses of several kilograys are sufficient to remove some hazardous pollutants, like pharmaceutical and personal care products, from contaminated water where the concentration of the pollutants is usually of the order of tens to hundreds μmol/L. It makes the process of radiation treatment of that kind promising for implementation.

3.3. Conversion of organic molecules and elimination of functional groups

Transformations in molecules of organic substance occurring under the action of active products of water radiolysis often play an important role in the purification of water and wastewater, even if TOC at that time changes slightly or does not change at all. It is of importance when the goal of "purification" is not to just remove organic compounds but to eliminate some unwanted properties of wastewater, like color, odor, toxicity, infection, or to increase the efficiency of further treatment in combined purification processes.

3.3.1. Disinfection

The process of polluted water/wastewater disinfection consists in the suppression of vital activity of microorganisms. Breaks occurred in the helix of DNA under direct or indirect (via radicals from water) action of radiation, destroying the ability of the cell to reproduce itself and resulting in microbe inactivation. As interaction of ionizing particle or active radical with any DNA molecule is an accidental event, the dependence of a number of living cells N per unit volume on absorbed dose D during irradiation is exponential:

$$N = N_0 \cdot e^{-\lambda D};\ \ln(N_0/N) = \lambda D,$$

where N_0 is the initial number of living cells and λ—inactivation constant depending on the kind of microorganism and irradiation conditions. In practice, to assess the efficiency of radiation disinfection, instead of λ, the value D_{10} is used which means the dose demanded to reduce the initial concentration of living cells by a factor of 10 and relates to λ as: $D_{10} = \ln 10/\lambda \cong 2.3/\lambda$.

Database [33] contains more than 1000 values of D_{10} measured to date. Typical values of D_{10} for different types of microorganisms are in the range of 0.4–1.3 kGy. That is why radiation sterilization of medical equipment, cosmetics, and so on is widely used, becoming a developed branch of industry. The application of radiation treatment for the sterilization of water and wastewater is undoubtedly effective, rather easily realized and has no objections, but one: high costs of ionizing radiation sources, which restrain its implementation at present.

3.3.2. *Detoxification*

Toxicity of many organic compounds is attributable to any specific functional groups, for example, halogenide-, nitrate-, sulfide groups, and some other, attached to hydrocarbon moiety, elimination of which decreases the toxicity sharply. Among the most widespread toxic pollutions, there are chlorine derivatives, the major constituent of "red list substances" [17]. Also, nitro groups attached to aliphatic or aromatic compounds increase the toxicity of the compound. One of the most effective radiation-induced processes of such functional groups detachment is the so-called dissociative electron attachment with the participation of hydrated electron or hydrogen atom:

$$\text{R–Cl} + e^-_{aq} \longrightarrow \text{R}^\bullet + \text{Cl}^-, \text{ or R–Cl} + {}^\bullet\text{H} \longrightarrow \text{R}^\bullet + \text{Cl}^- + \text{H}^+$$

A clear example of such process is consecutive dechlorination of pentachlorobiphenyl [34] which occurs under reductive conditions, that is, in the absence of dissolved oxygen and other scavengers of e^-_{aq} and $^\bullet H$. The yield of the process is close to summarized yield of e^-_{aq} and $^\bullet H$, that is, rather high. A similar process takes place in organic sulfides and nitro compounds:

$$\text{R–SH} + e^-_{aq} \longrightarrow \text{R}^\bullet + \text{SH}^-, \text{ or R–SH} + {}^\bullet\text{H} \longrightarrow \text{R}^\bullet + \text{SH}^- + \text{H}^+$$

$$\text{R–NO}_2 + e^-_{aq} \longrightarrow \text{R}^\bullet + \text{NO}_2^-, \text{ or R–NO}_2 + {}^\bullet\text{H} \longrightarrow \text{R}^\bullet + \text{NO}_2^- + \text{H}^+$$

The yield of the latter is lower because of the longer life time of intermediate radical anions $R\text{–}NO_2^{\bullet -}$ which partially disproportionate. Lower yield is also characteristic of dechlorination by $^\bullet OH$ radicals which, nevertheless, does occur [35].

It is necessary to keep in mind, however, that besides elimination of toxic groups, conversion of composite organic compounds upon irradiation may be realized in the formation of some products, toxicity of which is higher than that of initial compound [36]. Thus, under the action of ionizing radiation on clofibric acid solutions the total toxicity of the solutions at the beginning of irradiation sharply increases with increasing the dose and then decreases together with a decrease in the total content of organic matter [35]. Possible occurrence of two opposite radiolytic processes: detoxification of initial compound and formation of new toxic compounds, requires, when using radiation detoxification, taking into account real composition of water/wastewater to be treated in every particular case.

3.3.3. Decolorization

Many kinds of wastewater, especially that of textile and dyeing industry, are colored due to contamination with organic dyes. Composite molecules of dyes include chromophore group, that is, system of conjugated double bonds, responsible for the ability of the molecule to absorb electromagnetic radiation in the visible region of spectrum, that is, light. It is enough to destroy just chromophore group, not a whole molecule, to make the molecule incapable of light absorption and, thereby, to decolorize the dye. Hydrated electrons, as well as radicals $^{\bullet}H$ and $^{\bullet}OH$, are easily attached to a double bond (see subdivision 3.1.6) destroying the conjugated bonds system

$$e^{-}_{aq} + {>}C{=}C{<} \; (+H^{+}) \longrightarrow {>}CH{-}^{\bullet}C{<}$$

$$^{\bullet}H + {>}C{=}C{<} \longrightarrow {>}CH{-}^{\bullet}C{<}$$

$$^{\bullet}OH + {>}C{=}C{<} \longrightarrow {>}C(OH){-}^{\bullet}C{<}$$

The yield of decolorization, if only those reactions occurred, should be as high as 0.6 μmol/J corresponding to the total yield of all initial radicals from water. However, the occurrence of reduction-oxidation reactions of the latter resulted in the formation of two kinds of dye radicals: one in reaction of e^{-}_{aq} and $^{\bullet}H$, another in reaction of $^{\bullet}OH$, makes it possible, by electron and/or hydrogen transfer between two kinds of dye radicals, partial reparation of dye molecules. As a result, decolorization yields for aqueous solutions of soluble organic dyes, as a rule, are equal to 0.1–0.3 μmol/J, depending on the structure of dye (for disperse dyes, which are water-insoluble and form aqueous dispersions, the yield is lower because of impediments to radical penetration into dye particulates) [37, 38]. The yield of decolorization decreases with increasing absorbed dose, since the rate constants of radicals' reactions with original dye molecules and with the products of dye degradation are of the same order; so while the concentration of the latter increases during irradiation, the fraction of radicals being spent in reactions with those products also increases, reducing thereby the fraction of radicals required for "decolorization" of the original dye molecules.

In real wastewater, which contains, besides dye, rather high amounts of organic matter, decolorization yield is decreased due to competition of other organics for reactive radicals from water. Nevertheless, recently published results on radiation treatment of real wastewater from dyeing and textile industry [39] show that doses of several kilograys provide for sufficient decolorization of the wastewater.

3.3.4. Improvement of biological treatment

Biological treatment of wastewater with activated sludge is still the basic commercially applied method applied for the removal of great amount of organic matter from wastewater. However, many organic compounds which wastewater often contains are not biodegradable, that is, are not "eatable" by microorganisms of sludge. Moreover, some compounds are toxic for those microorganisms, "killing" them and, thereby, reducing the activity of sludge in biodegradation process.

Radiation-induced elimination of functional groups, as well as other kinds of molecule fragmentation, makes non-biodegradable organic compounds biodegradable and increases biodegradability of poorly biodegradable compounds. Also, the decomposition of toxic for microorganism compounds under irradiation improves biological treatment. Recently published results [40, 41] demonstrated that upon irradiation an increase in biodegradability of organic matter was observed without a significant decrease in the Total Organic Carbon. However, it should be mentioned that radiation treatment may have insignificant effect on biodegradability of wastewater with high amount of biodegradable organic matters as it has been observed on real textile and dyeing wastewater [42]. In that case, radiation treatment was much more effective when applied after biological treatment.

3.3.5. *Improvement of settling and sorption*

A combination of organic radicals formed in organic compound reaction with any active radicals from water results in the formation of more complex molecules. Due to further increase in molecules' size, the formation of solid phase becomes possible, especially, it is characteristic of natural organic matter. The formation and aggregation of small particulates of organic matter facilitates subsequent processes of sedimentation, filtration, and sorption. Thus, it has been found [43] that electron-beam treatment of river water contaminated, mainly, with NOM, significantly decreases the color index and the amount of suspended solids and improves water quality. A dose of several kilograys was found to be sufficient for decreasing the color index to a maximum permissible value even in the most unfavorable seasons (spring and autumn). Moreover, irradiation accelerated the sedimentation of suspended solids by a factor of 50–100, as compared with non-irradiated water. The precipitate formed upon irradiation captures impurities present in water; this process along with the radiolytic decomposition of impurities resulted in water purification.

3.4. Radiation polymerization

Polymerization of monomers', or oligomers', molecules in aqueous solutions under the action of ionizing radiation is initiated by primary products of water radiolysis, both radicals and ions [44, 45]. Accordingly, the process of polymerization may correspond to either a free-radical or an ionic (cationic or anionic) mechanism. Since primary anion-radicals and cation-radicals, as well as carbanions and carbocations, are very unstable in aqueous media, free-radical mechanism of polymerization usually prevails. It includes stages of

chain initiation (reaction of primary radical with a monomer molecule M).

$$^{\bullet}R + M \rightarrow {}^{\bullet}M';$$

chain propagation (increase in chain length):

$$^{\bullet}M' + M \rightarrow {}^{\bullet}M'M; \ldots {}^{\bullet}M'M_n + M \rightarrow {}^{\bullet}M'M_{n+1}$$

chain termination (radicals disappearance by combination or disproportionation).

$$^{\bullet}M'M_n + {}^{\bullet}M'M_n\ (\text{or}\,{}^{\bullet}R) \rightarrow \text{products;}$$

the polymerization rate during irradiation being proportional to the square root of dose rate and average polymer chain length—inversely proportional to the square root of dose rate.

The conversion of soluble or slightly soluble molecules of monomers/oligomers into insoluble polymer upon irradiation makes it possible to apply the process of radiation polymerization to the removal of monomer impurities which are often present in wastewater of chemical and textile industries. However, because of a decrease in polymer length with increasing the dose rate using electron-beam irradiation of high intensity reduces the effectiveness of the process. Nevertheless, combined radiation-induced processes of decomposition and polymerization were shown to be quite effective in the purification of aqueous systems from such typical refractory organic pollutant as polyvinyl alcohol [46].

3.5. Oxidation-reduction processes in inorganic compound solutions

The application of radiation treatment to change mineral composition of water or wastewater is rarely considered to be an efficient one. Nevertheless, several radiation induces reductive-oxidative reaction of inorganic compounds may be used in the processes of inorganic toxic compounds removal.

Radicals $^{\bullet}OH$ possessing high oxidation potential oxidizes all transition metal cations in lower oxidation state and majority of inorganic anions (see subdivision 3.1.6.3". It may be applied to convert, for example, toxic nitrite anions present in water/wastewater into less toxic nitrate anions:

$$^{\bullet}OH + NO_2^- \rightarrow OH^- + NO_2^{\bullet}; 2NO_2^{\bullet}\ (+H_2O) \rightarrow NO_3^- + NO_2^-\ (+2H^+)$$

Hydrated electrons and hydrogen atoms as strong reducing agents reduce majority of metal cations and oxygen-containing anions (see subdivisions 3.1.6.1 and 3.1.6.2). That process was tested in the conversion of highly toxic Cr(VI) ions into much less toxic Cr(III) ions [47] in reactions:

$$e^-_{aq}\ (^{\bullet}H) + Cr(VI) \rightarrow Cr(V)\ (+H^+); 2Cr(V) \rightarrow Cr(VI) + Cr(IV); 2\ Cr(IV) \rightarrow Cr(V) + Cr(III)$$

as well as in the removal of heavy metals, like cadmium [48] and lead [49] from aqueous system by reducing metal cations to metal atoms with subsequent sedimentation of solid phase and filtration or sorption:

$$e^-_{aq}\ (^{\bullet}H) + Cd(II) \rightarrow Cd(I)\ (+H^+); 2Cd(I) \rightarrow Cd(II) + Cd(0); nCd(0) \rightarrow Cd_n\downarrow$$

Since $^{\bullet}OH$ radicals oxidize reduced cations backward, the process becomes efficient in the presence of $^{\bullet}OH$ radicals scavengers, like formates, sulfides, sulfites, and so on. Combined radiation-flotation process based on radiation-induced reductive-oxidative reaction of Hg has been implemented to remove mercury from wastewater of hydrolysis works [50].

3.6. Processes of micelles recharging in aqueous dispersions

Aqueous dispersions are very sensitive to the action of ionizing radiation, which may either increase stability of such systems or initiate coagulation even at rather low-absorbed doses.

One of the main reasons for coagulation is recharging the micelles by interaction with short-lived products of water radiolysis. Thus, negatively charged micelles preferably interact with positively charged or uncharged intermediates, participating, for example, in reactions of protonation by H^+ ions or electron transfer to $^{\bullet}OH$ and $HO_2^{\bullet}$ radical:

$$Mic^{n-} + H^+ \rightarrow MicH^{(n-1)-}$$

$$Mic^{n-} + {}^{\bullet}OH \rightarrow Mic^{(n-1)-} + OH^-$$

$$Mic^{n-} + HO_2^{\bullet} \rightarrow Mic^{(n-1)-} + HO_2^-$$

Appearing misbalance in charges of nucleus and outer shell of micelle can promote consecutive combination with other micelles:

$$m\ Mic_m{}^{n-} + Mic^{(n-1)-} \rightarrow Mic_{m+1}{}^{(m+n-1)-}$$

and, finally, the formation of large aggregates settling down from the solution. A similar situation takes place with positively charged micelles, in which positive outer-shell charge is reduced by interaction with hydrated electrons and/or hydrogen atoms:

$$Mic^{n+} + e^-{}_{aq}\ ({}^{\bullet}H) \rightarrow MicH^{(n-1)+}\ (+H^+)$$

There exists, however, another source of influence on charged micelles, which is only characteristic of electron-beam irradiation unlike irradiation with gamma or X-rays. The most distinguishing feature of irradiation by electrons under condition of their complete absorption in irradiated medium consists in the accumulation of uncompensated charge due to incident electrons. These electrons after dissipation of their energy in the processes of ionization (thereby, equal quantities of oppositely charged particles arise) and excitation are finally thermalized and localized in aqueous medium giving rise to the formation of $e^-{}_{aq}$, which carry excess negative charge. Fraction of $e^-{}_{aq}$ formed by deceleration of incident electrons is low relative to $e^-{}_{aq}$ formed by ionization (for 0.7 MeV electrons it is about 0.05%), and their role in radiation-induced oxidation-reduction processes is negligible. However, while $e^-{}_{aq}$ themselves have very short lifetime because of high reactivity, additional charge can be eliminated only by diffusion of anions to vial walls and discharging into environment, which courses relatively slow. Therefore, stationary concentration of uncompensated negative charge during irradiation may be sufficiently high to have an influence on coagulation process by stabilizing negative micelles. It was observed recently [51] in experiments concerning coagulation of humic acids and lignin in aqueous dispersions.

4. Ionizing radiation sources for radiation technologies of aqueous system treatment

Ionizing radiation includes electromagnetic radiation of wavelength shorter than 10^{-7} m and accelerated ions with kinetic energy higher than lowest ionization energy of molecules, that is, near 10 eV (neutron radiation is often referred to as ionizing one, but neutrons themselves do not interact with electron shells of molecules, ionization by neutrons being indirect). All the sources of such radiation may be divided into two principal groups: radioactive isotopes (radionuclides) and accelerating techniques. With the help of the latter, both ion beams and electromagnetic radiation (X-rays) may be generated. All the ionizing radiation sources intended for commercial application must meet the following major requirements: be safe for environment, provide for sufficient depth of irradiation, and absorbed dose. Now, three kinds of sources are applied in radiation processing including radiation treatment of aqueous systems: radioisotope sources, X-ray generators, and electron accelerators.

4.1. Isotope sources

Radiation sources based on only two γ-emitting isotopes were used in radiation technologies. There are radioisotopes of cesium (Cs-137) and cobalt (Co-60).

Radioisotope Cs-137 is produced by separation from other fission products formed in nuclear reactors. It decays with a half-life of 30.17 years, emitting beta-rays with energy of 0.51 MeV and gamma-rays with energy of 0.667 MeV; the product of Cs-137 decay is stable isotope of barium (Ba-137). In radiation sources isotope Cs-137 is used in the form of cesium chloride salt encapsulated in a metal shell.

Radioisotope Co-60 is produced in nuclear reactors by bombardment of stable isotope Co-59 with neutrons. It decays with a half-life of 5.27 years, emitting beta-rays with energy of 0.32 MeV and gamma-rays with energies of 1.173 and 1.332 MeV (mean gamma-rays energy of 1.25 MeV); the product of Co-60 decay is a stable isotope of nickel (Ni-60). In radiation sources, radioisotope Co-60 is used in the form of metal slugs or pellets.

While penetrating a medium, gamma radiation is attenuated exponentially:

$$I = I_0 \cdot \mathrm{e}^{-\mu l}; \lg(I_0/I) = \mu l; l_{1/2} = \lg 2/\mu \cong 0.7/\mu$$

where I is the intensity of gamma radiation (I_0—initial intensity), l—passed length ($l_{1/2}$—length of half-attenuation), and μ—linear coefficient of attenuation which depends on radiation quanta energy and on medium density. In aqueous systems, values of $l_{1/2}$ for cesium and cobalt gamma-radiations equal, respectively, 8 and 11 cm [19], differing due to difference in the energy of radiation.

Because of high penetration of gamma-rays gamma-emitting sources must be shielded from environment with a sufficient thick wall of dense material which is usually lead (in compact sources) or concrete (normally near 2 m in thickness, in industrial irradiation facilities). Major components of an industrial facility include irradiation room, shielded storage room (dry or wet) for the radiation source rack, source hoist mechanism, radiation shield surrounding the irradiation room, control console (room), feed system for a product to be irradiated, shielding maze, control and safety interlock system, and supporting service equipment [52].

When being used in the radiation treatment of aqueous systems, gamma sources, although, meet the requirement of irradiation depth, possess significant disadvantages:

- Radiation from the source is emitted spherically in all directions, so a part of it is wasted, reducing thereby power efficiency of the source.

- Radiation from the source is emitted continuously with constant intensity, so it is not easy to switch off and on irradiation promptly and to change the dose rate.
- Gamma sources have relatively low power; thus, the source of such high activity as 1 MCi (3.7×10^{16} Bq) has a power just near 10 kW. It makes gamma sources inapplicable for the treatment of aqueous systems at high flow rates.

Nowadays, Co-60 gamma sources are applied, as a rule, to reveal the mechanisms of processes in systems to be irradiated, to disinfect sewage sludge, and to develop technologies of some processes. Cs-137 gamma sources are presently not used for solving environmental problems including radiation treatment of water/wastewater.

4.2. X-ray sources

Both γ-rays and X-rays are electromagnetic radiation and differ by the principle of formation. While γ-rays are emitted in the process of energy transition from excited to ground state of a nucleus, X-rays are emitted by accelerated electrons when they pass near to atomic nuclei and are deflected by their electric fields, the effect being called bremsstrahlung. Contrary to γ-rays with defined energy, X-rays produced in this manner have a continuous spectrum of photon energies extending from about 30 keV up to a maximum energy equal to the kinetic energy of the incident electron. For example, with a 5-MeV electron, the most probable photon energy is about 300 keV, the average photon energy is near 1.0 MeV, while the maximum photon energy is 5.0 MeV. The X-ray intensity increases with the electron energy and beam current and with the atomic number of the target material [53].

Two kinds of X-ray sources are produced and used commercially: X-ray tubes (or Roentgen tubes) and X-ray generators, based on electron accelerators. The former have relatively low photons energy and low power. Because of this X-ray tubes are almost not applied in radiation processing, being widely applied in medicine, material and structure analysis, and for industrial inspection.

X-ray generators are powerful analog of gamma sources. A comparison of those two kinds of electromagnetic radiation sources has been made in [54]. A distinguishing characteristic of X-ray generators is ameliorating the parameters of generated electromagnetic radiation with increasing the energy of incident electrons. Thus, while at relatively low electron energies X-rays are emitted in all directions, with electron energies higher than 2.0 MeV the X-ray emission in the direction of the incident electron beam becomes prevailing. Further increase in electrons energy results in decreasing the angular dispersion of the X-ray beam: divergence measured at half of the maximum intensity (HMI) is reduced two times when increasing electrons energy from 5 to 10 MeV (see **Table 9**).

Electron energy (MeV)	Mean photon energy (MeV)	Emission efficiency (%)	Divergence at HMI (degrees)
5.0	1.19	8.2	20
7.5	1.38	13.3	15
10	1.56	16.2	10

Table 9. Parameters of generated electromagnetic radiation depending on electron beam energy (the data from [55]).

Also, both efficiency of electron beam conversion into X-ray beam and photon energy of X-rays are increased with increasing the energy of accelerated electrons. However, electron accelerators with energy higher than 7.5 MeV are forbidden to application in industry because of the danger of activating the matter under irradiation.

While being a prosperous competitor to radioisotope sources in many kinds of radiation processing (recent comparisons have shown that the capital costs and electric power costs for accelerators with electron energies of 5.0–7.0 MeV can be lower than the capital costs and source replenishment costs for cobalt-60 source loadings greater than 2.0 MCi [55]), X-ray generators, nevertheless, are not suitable for aqueous systems treatment. The reason is that in spite of capability to irradiate rather a thick layer of condensed media, the energy efficiency of X-ray generators is very low. At present, they are not proposed to be used in the field of environment conservation.

4.3. Electron accelerators

Among all accelerators of charged particles, only accelerators of electrons are suitable to application in radiation processing (excluding ion accelerators used in ion implantation industries). Chemical consequences of accelerated electrons action on many substances, including aqueous media, are almost the same as those of gamma and X-rays. The difference is in the depth of penetration. While the intensity of electromagnetic radiation (both gamma and X-rays) in a medium is attenuated exponentially with path length, accelerated electrons have limited penetration depth which depends on their energy. In **Table 10**, the penetration of gamma radiation is compared with that of accelerated electrons of different energies.

Principal characteristics of electron accelerators, important for industrial application, are energy of accelerated electrons E_e (or "accelerator energy") and beam power W_e, which is defined as E_e multiplied by electron beam current I_e: $W_e = E_e{\cdot}I_e$. While E_e determines the range of electrons in irradiated medium, that is, the depth of electron beam penetration, W_e determines the productivity of irradiation process. Market cost of accelerators is near proportional to their energy and to the square root of power. It is accepted in radiation processing to divide electron accelerators according to accelerator energy into three groups: 1—low-energy units (E_e < 0.5 MeV), 2—mid-energy units (0.5 MeV < E_e < 5.0 MeV), and 3—high-energy units (5.0 < E_e < 10 MeV), maximum energy of 10 MeV being set for industrial accelerators to avoid problems of induced radioactivity.

Two types of accelerator design have found industrial acceptance in the high-energy area (5.0–10 MeV): microwave linear accelerators (linacs) and the radiofrequency accelerator Rhodotron®. Now industrial linacs are produced by companies: Getinge Linac (formerly Linac Technologies), Budker Institute of Nuclear Physics, L-3 Communications Pulse

Gamma radiation		Accelerated electrons				
Cs-137	Co-60	0.5 MeV	1 MeV	2 MeV	5 MeV	10 MeV
8	11	0.2	0.4	0.8	2.3	4.7

Table 10. Depth of half attenuation for gamma radiation and penetration depth for accelerated electrons in cm (the data from [19]).

Sciences (formerly Titan Scan), Mevex, and EB-Tech Co Ltd. (in cooperation with the Budker Institute of Nuclear Physics); Rhodotron®—by Ion Beam Applications SA (formerly Radiation Dynamics Inc) [56].

While being rather widely used in other fields of radiation processing, industrial high-energy accelerators are not used in the radiation treatment of water and wastewater, mainly due to high cost (including the cost of accelerator itself and the necessity of strong biological shielding because of high-intensity level of induced X-ray radiation) and relatively low-energy efficiency of linacs (ratio of beam power to power consumption is about 30%).

Low-energy accelerators produced by companies: Energy Sciences Incorporated, Broadbeam Equipment, NHV Corporation, and Advanced Electron Beams, are used, mainly, for curing or crosslinking of inks, coatings, and adhesives that are based on liquid-reactive materials which do not contain solvents. Because of very thin layer of water which electrons generated by those accelerators can penetrate, they are not applicable in radiation treatment of aqueous systems but one case; irradiation of sprayed wastewater (see Section 5.1).

Mid-energy electron accelerators are the most suitable for aqueous systems treatment. Those accelerators produce scanned beams that range in energy from 400 to 5 MeV. The units that are suitable for industrial use are all characterized by the ability to produce high beam currents, many tens of milliamps, which provide high dose rates that are needed for the treatment of aqueous systems at high flow rates.

Five electrical design systems have been used to attain mid-energy and high beam current: 1—the Cockcroft-Walton and its enhancements by Nissin-High Voltage, 2—the Insulated Core Transformer, 3—the Dynamitron, 4—a magnetic-coupled dc system, and 5—high-current pulsed beams. The Dynamitron can attain very high beam currents (60 mA) at up to 5.0 MeV (300 kW). The ICT and ELV mid-energy accelerator designs are limited in electron energy to 2.5 MeV. Medium- and low-voltage direct current accelerators may have several accelerating tubes connected to one power supply by using cable connections or gas-insulated transmission lines, flexible cable connections being limited to 800 keV and gas transmission systems up to 2 MeV. The companies that have experience in providing industry with mid-energy, high-current electron accelerators are Ion Beam Applications SA (formerly Radiation Dynamics Inc), NHV Corporation, Budker Institute of Nuclear Physics, Efremov Scientific Research Institute of Electrophysical Apparatus (NIIEFA), Vivirad SA (formerly Vivirad-High Voltage Corporation), and Wasik Associates [56].

As mentioned earlier, the cost of accelerators of similar energy is near proportional to the square root of power. It means that the cost per unit power decreases with increasing the power of accelerator. The data cited in [57] show that typical prices of mid-energy accelerators with power of 20–400 kW increase from 0.5 to 2.5 $million, while the cost per 1 kW decreases from $25,000 to $5000.

5. Under-beam and other support equipment

Besides an accelerator, support equipment plays an important role in realizing radiation technology of water/wastewater treatment. The equipment includes radiation reactor, as a main part, and other equipment necessary for normal functioning of radiation treatment facility.

5.1. Reactor types: systems of water feed

Radiation reactor is an installation where flow of water is irradiated. There exist three different methods to irradiate water flow. Conditionally, they may be called "natural flow method," "spray method," and "wide jet method," the reactors being called correspondingly.

5.1.1. Natural flow reactor

In this method, water/wastewater either flows over a surface and is irradiated from above by a vertical electron beam (**Figure 2a**), or falls as a wide waterfall and is irradiated by a horizontal electron beam (**Figure 2b**).

Water/wastewater treatment in this method may be carried out at high doses and dose rates. However, the velocity of natural flow is limited by about 1 m/s. Since the thickness of water layer is limited by accelerated electrons energy, the throughput of irradiation facility may be increased only by an increase in flow width, which results in increasing irradiation area and demand of additional accelerator (or additional accelerating tube).

The advantage of the method consists in a possibility of equipping the surface under flow with holes to input ozone or other necessary additions and/or to mix liquid under irradiation by gas bubbling. Reactors of that type were used in many pilot plans on radiation treatment of water, wastewater, and sludge, in USA, Brazil, European countries, Korea, Japan, and China.

5.1.2. Spray reactor

In this method, polluted water is sprayed under pressure into a chamber to be irradiated from above by a vertical electron beam (**Figure 3**).

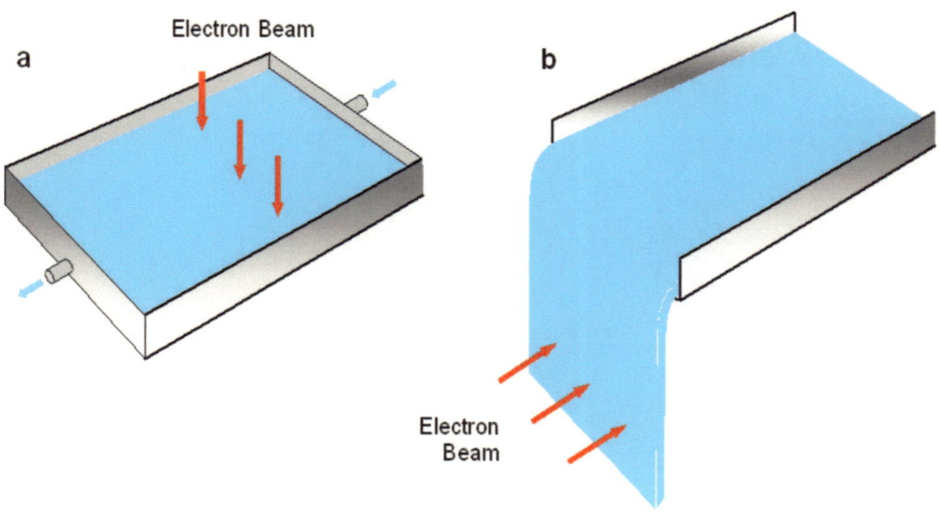

Figure 2. Natural flow reactor with vertical (a) and horizontal (b) direction of electron-beam irradiation.

While flow velocity is up to 3 m/s, the throughput of such irradiation facility is low. Also, the water before being input into reactor should be thoroughly purified from suspended particulates to prevent injector clogging and damaging.

An advantage of the method consists in the formation of large specific surface of sprayed liquid providing for efficient penetration of surrounding gas into liquid phase. If the gas is an air or oxygen, ozone being formed under irradiation and absorbed in solution takes part in radiation-chemical reactions which improve purification process; to create reductive conditions during irradiation, the chamber may be filled with inert gas.

5.1.3. Wide jet reactor

In this method, wastewater is fed under pressure to irradiation area in the form of wide jet and is irradiated from above by a vertical electron beam (**Figure 4**).

Since flow velocity in this method may be made several meters per second, this kind of reactors is the most promising to irradiate wastewater using powerful accelerators at high flow rates. It was employed in full-scale facility in Korea on radiation treatment of dyeing company wastewater with throughput of 10,000 m^3 a day [58].

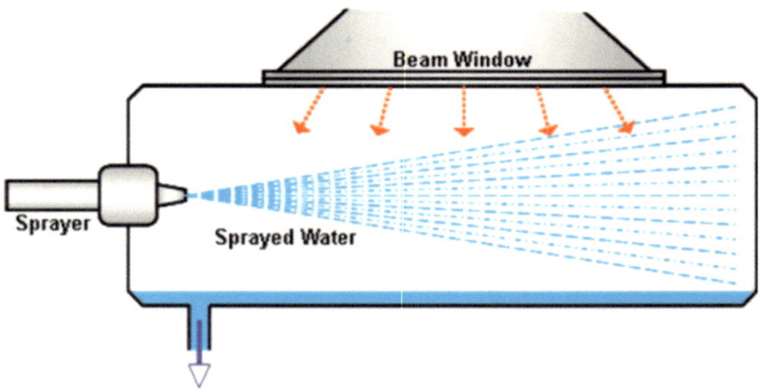

Figure 3. Spray reactor.

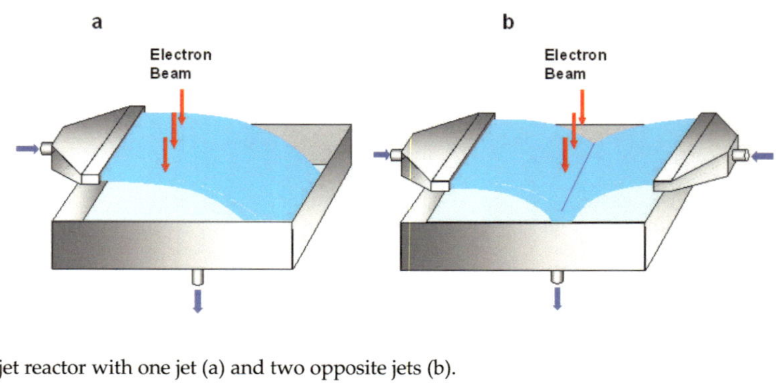

Figure 4. Wide jet reactor with one jet (a) and two opposite jets (b).

5.2. Other equipment

In addition to an accelerator and a reactor, other equipment is necessary to provide functioning the facility on radiation treatment of water/wastewater. It includes

- Tubing for inlet and outlet water;
- Water pumps, some of which in the case of wide jet reactor should be of high pressure and capacity;
- Air fans (blowers) for cooling electron beam window of accelerator, discharge of ozone containing air from irradiation room, ventilation;
- Remote control to operate an accelerator;
- Chemical laboratory equipment to analyze inlet and outlet water;
- Measurement instrumentation, and so on. [58].

The cost of reactor and listed equipment may amount to one-third of accelerator cost.

Besides, the necessary part of the facility is biological shielding. Powerful accelerators must be situated in special irradiation room surrounded by walls made of dense concrete of 1.5–2.0 m thickness or more. The cost of the wall building amounts up to half of the accelerator cost. Thus, the capital cost of an industrial plant for wastewater treatment of middle productivity is near 4.0 $million, excluding costs of land, R&D, and approval from authorities [59]. The demand of such high investments is one of the main obstacles in developing and implementing radiation technologies of aqueous systems purification.

6. Conclusion

Radiation treatment of aqueous systems may be effectively used, without additional purification methods, for disinfection, detoxification, and decolorization of contaminated water at low concentration of organic matter. Purification of wastewater with high concentration of organic matter requires the application of combined radiation method, including biological treatment and/or filtration, sorption, and so on. Radiation technologies of aqueous systems treatment at present are being extensively developed. Electron accelerators are preferable radiation sources, especially at high flow rates. One of prerequisites for expansion of the technologies should be reducing the market prices for radiation equipment, namely electron accelerators.

Author details

Igor E. Makarov* and Alexander V. Ponomarev

*Address all correspondence to: makarov@ipc.rssi.ru

Frumkin Institute of Physical Chemistry and Electrochemistry, Russian Academy of Sciences, Moscow, Russia

References

[1] Woods RJ, Pikaev AK. Applied Radiation Chemistry: Radiation Processing. New York: Wiley; 1994. p. 535

[2] Berejka AJ, Cleland MR, editors. Industrial Radiation Processing with Electron Beams and X-Rays. Vienna: IAEA & IIA; 2011. p. 105

[3] Parsons S, editor. Advanced Oxidation Processes for Water and Wastewater Treatment. London: IWA Publishing; 2004. p. 372

[4] Stefan MI, editor. Advanced Oxidation Processes for Water Treatment: Fundamentals and Applications. London: IWA Publishing; 2017. p. 450

[5] Wojnárovits L, Takács E. Wastewater treatment with ionizing radiation. Journal of Radioanalytical and Nuclear Chemistry. 2017;**311**:973-981. DOI: 10.1007/s10967-016-4869-3

[6] von Sperling M. Biological Wastewater Treatment Series. Vol. 1. Wastewater Characteristics, Treatment and Disposal. London: IWA Publishing; 2007. p. 308

[7] Uyttendaele M, Jaykus L-A, Amoah P, Chiodini A, Cunliffe D, Jacxsens L, Holvoet K, Korsten L, Lau M, McClure P, Medema G, Sampers I, Jasti PR. Microbial hazards in irrigation water: Standards, norms, and testing to manage use of water in fresh produce primary production. Comprehensive Reviews in Food Science and Food Safety. 2015;**14**:336-356. DOI: 10.1111/1541-4337.12133

[8] Connor R. The global perspective on water. In: The United Nations World Water Development Report 2016: Water and Jobs. Paris: UNESCO; 2016. p. 15-29. ISBN: 978-92-3-100155-0

[9] Schmoll O, Howard G, Chilton G, Chorus I, editors. Protecting Groundwater for Health: Managing the Quality of Drinking-Water Sources. London: IWA Publishing; 2006. p. 697

[10] Rickert B, Chorus I, Schmoll O. Protecting Surface Water for Health. Identifying, Assessing and Managing Drinking-Water Quality Risks in Surface-Water Catchments. Geneva: WHO; 2016. p. 196

[11] Guidelines for Drinking-Water Quality. Third edition incorporating 1st and 2nd addenda, Vol.1, Recommendations. Geneva: WHO; 2008. p. 515

[12] Pharmaceuticals in Drinking-Water. Geneva: WHO; 2012. p. 35

[13] Potable Reuse: Guidance for Producing Safe Drinking-Water. Geneva: WHO; 2017. p. 138

[14] Chemical Mixtures in Source Water and Drinking-Water. Geneva: WHO; 2017. p. 104

[15] Guidelines for Drinking-Water Quality: Fourth Edition Incorporating the First Addendum. Geneva: WHO; 2017. p. 541

[16] Olivieri AW, Crook J, Anderson MA, Bull RJ, Drewes JE, Haas CN, Jakubowski W, McCarty PL, Nelson KL, Rose JB, Sedlak DL, Wade TJ. Expert Panel Final Report: Evaluation

of the Feasibility of Developing Uniform Water Recycling Criteria for Direct Potable Reuse. Prepared August 2016 by the National Water Research Institute for the State Water Resources Control Board, Sacramento. p. 382. Available from: https://books.google.nl/books?id=m5mYAQAACAAJ

[17] The United Nations World Water Development Report. Wastewater: The Untapped Resource. Paris: UNESCO; 2017. p. 180

[18] Hanjra MA, Blackwell J, Carr G, Zhang F, Jackson TM. Wastewater irrigation and environmental health: Implications for water governance and public policy. International Journal of Hygiene and Environmental Health. 2012;**215**(3):255-269. DOI: 10.1016/j.ijheh.2011.10.003

[19] Spinks JWT, Woods RJ. An Introduction to Radiation Chemistry. New York: Wiley; 1990. p. 592

[20] Buxton GV, Greenstock CL, Helman WP, Ross AB. Critical review of rate constants for reactions of hydrated electrons, hydrogen atoms and hydroxyl radicals (·OH/$O^{-·}$) in aqueous solution. Journal of Physical and Chemical Reference Data. 1988;**17**:513-886. DOI: 10.1063/1.555805

[21] Bielski BH, Cabelli DE, Arudi RL, Ross AB. Reactivity of HO_2/O_2^- radicals in aqueous solution. Journal of Physical and Chemical Reference Data. 1985;**14**:1041-1100. DOI: 10.1063/1.555739

[22] Alkhuraiji TS, Leitner NKV. Effect of oxidant addition on the elimination of 2-naphthalenesulfonate in aqueous solutions by electron beam irradiation. Radiation Physics and Chemistry. 2016;**126**:95-102. DOI: 10.1016/j.radphyschem.2016.05.016

[23] Kovács K, He S, Míle V, Földes T, Pápai I, Takács E, Wojnárovits L. Ionizing radiation induced degradation of monuron in dilute aqueous solution. Radiation Physics and Chemistry. 2016;**124**:191-197. DOI: 10.1016/j.radphyschem.2015.10.028

[24] Alkhuraiji TS, Ajlouni A-W. Destruction of amphetamine in aqueous solution using gamma irradiation. Radiation Physics and Chemistry. 2017;**139**:17-44. DOI: 10.1016/j.radphyschem.2017.05.014

[25] Alkhuraiji TS, Boukari SOB, Alfadhl FS. Gamma irradiation-induced complete degradation and mineralization of phenol in aqueous solution: Effects of reagent. Journal of Hazardous Materials. 2017;**328**:29-36. DOI: 10.1016/j.jhazmat.2017.01.004

[26] Paul I, Naik DB, Bhardwaj YK, Varshney L. Studies on oxidative radiolysis of ibuprofen in presence of potassium persulfate. Radiation Physics and Chemistry. 2014;**100**:38-44. DOI: 10.1016/j.radphyschem.2014.03.016

[27] Kubesch K, Zona R, Solar S, Gehringer P. Degradation of catechol by ionizing radiation, ozone and the combined process ozone-electron-beam. Radiation Physics and Chemistry. 2005;**72**(4):447-453. DOI: 10.1016/j.radphyschem.2004.01.007

[28] Zheng BG, Zheng Z, Zhang JB, Luo XZ, Wang JQ, Liu Q, Wang LH. Degradation of the emerging contaminant ibuprofen in aqueous solution by gamma irradiation. Desalination. 2011;**276**:379-385. DOI: 10.1016/j.desal.2011.03.078

[29] Guo Z, Zhou F, Zhao Y, Zhang C, Liu F, Bao C, Lin M. Gamma irradiation-induced sulfadiazine degradation and its removal mechanisms. Chemical Engineering Journal. 2012;**191**:256-262. DOI: 10.1016/j.cej.2012.03.012

[30] Abdel daiem MM, Rivera-Utrilla J, Ocampo-Perez R, Sanchez-Polo M, Lopez-Penalver JJ. Treatment of water contaminated with diphenolic acid by gamma radiation in the presence of different compounds. Chemical Engineering Journal. 2013;**219**:371-379. DOI: 10.1016/j.cej.2012.12.069

[31] Chu L, Wang J. Degradation of 3-chloro-4-hydroxybenzoic acid in biological treated effluent by gamma irradiation. Radiation Physics and Chemistry. 2016;**119**:194-199. DOI: 10.1016/j.radphyschem.2015.10.016

[32] Wang J, Chu L. Irradiation treatment of pharmaceutical and personal care products (PPCPs) in water and wastewater: An overview. Radiation Physics and Chemistry. 2016;**125**:56-64. DOI: 10.1016/j.radphyschem.2016.03.012

[33] Micro D10 referencedata [Internet]. 2017. Available from: http://www.irradiationpanel.org/app/download/6497933/Micro_D10_referencedata.xls [Accessed: 3 September 2017]

[34] Ponomarev AV, Makarov IE, Kim Y, Han B. Formation and decay of biphenyl at consecutive radiolytic dechlorination of pentachlorobiphenyl in aqueous solution. Mendeleev Communication. 2005;**15**(6):246-248. DOI: 10.1070/MC2005v015n06ABEH002159

[35] Csay T, Rácz G, Salik Á, Takács E, Wojnárovits L. Reactions of clofibric acid with oxidative and reductive radicals—Products, mechanisms, efficiency and toxic effects. Radiation Physics and Chemistry. 2014;**102**:72-78. DOI: 10.1016/j.radphyschem.2014.04.019

[36] Borrely SI, Gonçalves AA, Oikawa H, Duarte C, Rocha FR. Electron beam accelerator for detoxification of effluents when radiation process can enhance acute toxic? Radiation Physics and Chemistry. 2004;**71**:455-458. DOI: 10.1016/j.radphyschem.2004.03.087

[37] Rauf MA, Ashraf SS. Radiation induced degradation of dyes—An overview. Journal of Hazardous Materials. 2009;**166**(1):6-16. DOI: 10.1016/j.jhazmat.2008.11.043

[38] Wojnárovits L, Takács E. Irradiation treatment of azo dye containing wastewater: An overview. Radiation Physics and Chemistry. 2008;**77**(3):225-244. DOI: 10.1016/j.radphyschem.2007.05.003

[39] Borrely SI, Morais AV, Rosa JM, Badaró-Pedroso C, Pereira MC, Higa MC. Decoloration and detoxification of effluents by ionizing radiation. Radiation Physics and Chemistry. 2016;**124**:198-202. DOI: 10.1016/j.radphyschem.2015.11.001

[40] Szabó L, Tóth T, Homlok R, Rácz G, Takács E, Wojnárovits L. Hydroxyl radical induced degradation of salicylates in aerated aqueous solution. Radiation Physics and Chemistry. 2014;**97**:239-245. DOI: 10.1016/j.radphyschem.2013.11.039

[41] Sági G, Kovács K, Bezsenyi A, Csay T, Takács E, Wojnárovits L. Enhancing the biological degradability of sulfamethoxazole by ionizing radiation treatment in aqueous solution. Radiation Physics and Chemistry. 2016;**124**:179-183. DOI: 10.1016/j.radphyschem.2016.02.009

[42] He S, Sun W, Wang J, Chen L, Zhang Y, Yu J. Enhancement of biodegradability of real textile and dyeing wastewater by electron beam irradiation. Radiation Physics and Chemistry. 2016;**124**:203-207. DOI: 10.1016/j.radphyschem.2015.11.033

[43] Pikaev AK, Bludenko AV, Makarov IE, Ponomarev AV, Minin VN, Ponomarev VI, Linnik OA. Electron-beam treatment of highly-colored river water. Radiation Physics and Chemistry. 1996;**48**(1):75-80. DOI: 10.1016/0969-806X(95)00425-W

[44] Rosiak JM. Radiation polymerization in solution. In: Advances in Radiation Chemistry of Polymers. IAEA-TECDOC-1420. Vienna: IAEA; 2003. pp. 41-60

[45] Dispenza C, Alessi S, Spadaro J. Radiation processing of polymers in aqueous media. In: Sun Y, Chmielewski AG, editors. Applications of Ionizing Radiation in Materials Processing. Warszawa: INCT; 2017. pp. 291-326

[46] Sun W, Chen L, Tian J, Wang J, He S. Radiation-induced decomposition and polymerization of polyvinyl alcohol in aqueous solutions. Environmental Engineering and Management Journal. 2013;**12**(7):1323-1328 ISSN: 1843-3707

[47] Yurik TK, Pikaev AK. Radiolysis of weakly acidic and neutral aqueous solutions of hexavalent chromium ions. High Energy Chemistry. 1999;**33**(4):208-212 ISSN: 0018-1439

[48] Kartasheva LI, Chulkov VN, Didenko OA, Pikaev AK, Kim DK, Kim Y, Han B. Cadmium removal from water by combined electron-beam and adsorption treatment. High Energy Chemistry. 1998;**32**(2):127-128 ISSN: 0018-1439

[49] Zhestkova TP, Pikaev AA, Pikaev AK, Kim DK, Kim Y, Han B. Lead removal from aqueous solutions by electron-beam treatment in the presence of a scavenger of OH radicals. High Energy Chemistry. 1997;**31**(4):222-225 ISSN: 0018-1439

[50] Lyu E, Makarov IE, Kuryatnikov YI, Mazanko RV, Dzhagatspanyan RV, Pikaev AK. Radiation-flotation purification of aqueous wastes from mercury. Radiation Physics and Chemistry. 1983;**22**(3–5):503-509. DOI: 10.1016/0146-5724(83)90057-2

[51] Makarov IE, Metreveli PK, Metreveli AK, Ponomarev AV. Effect of electron irradiation on dispersions of humic acids and lignin. Radiation Physics and Chemistry. 2014;**97**:12-15. DOI: 10.1016/j.radphyschem.2013.10.010

[52] Gamma Irradiators for Radiation Processing. Vienna: IAEA; 2005. p. 40 Available from: www.naweb.iaea.org/napc/iachem/Brochgammairradd.pdf

[53] Seltzer SM, Farrell JP, Silverman J. Bremsstrahlung beams from high-power electron accelerators for use in radiation processing. IEEE Transactions on Nuclear Science. 1983; **NS-30**(2):1629-1633. DOI: 10.1109/TNS.1983.4332602

[54] Cleland MR, Pageau GM. Comparisons of X-ray and gamma-ray sources for industrial irradiation processes. Nuclear Instruments and Methods in Physics Research B. 1987; **24–25**(2):967-972. DOI: 10.1016/S0168-583X(87)80290-2

[55] Cleland MR, Stichelbaut F. Radiation processing with high-energy X-rays. In: 2009 International Nuclear Atlantic Conference—INAC; 27 September to 2 October 2009; Rio de Janeiro. ISBN: 978-85-99141-03-8. Available from: http://www.iaea.org/inis/collection/NCLCollectionStore/_Public/41/097/41097063.pdf

[56] Industrial radiation processing with electron beams and X-rays. Vienna: IAEA; 2011. p. 105

[57] Radiation Processing: Environmental Application. Vienna: IAEA; 2007. p. 71

[58] Makarov IE, Ponomarev AV, Han B. Demonstration plant for electron-beam treatment of Taegu dye industry complex wastewater. In: Emerging Application of Radiation Processing. (IAEA TechDoc-1386). Vienna: IAEA; 2004. pp. 138-152

[59] Han B, Kim JK, Kim Y, Choi JS, Jeong KY. Operation of industrial-scale electron beam wastewater treatment plant. Radiation Physics and Chemistry. 2012;**81**(9):1475-1478. DOI: 10.1016/j.radphyschem.2012.01.030

Chapter 6

Ionizing Radiation-Induced Polymerization

Mohamed Mohamady Ghobashy

Additional information is available at the end of the chapter

http://dx.doi.org/10.5772/intechopen.73234

Abstract

Ionizing radiation can induce some kinds of reactions, other than polymerization, such as dimerization, oligomerization, curing, and grafting. These reactions occur through a regular radical chain causing growth of polymer by three steps, namely, initiation, propagation, and termination. To understand ionizing radiation-induced polymerization, the water radiolysis must be taken into consideration. This chapter explores the mechanism of water molecules radiolysis paying especial attention to the basic regularities of solvent radicals' interaction with the polymer molecules for forming the crosslinked polymer. Water radiolysis is the main engine of the polymerization processes, especially the "free-radical polymerization." The mechanisms of the free-radical polymerization and crosslinking will be discussed in detail later. Since different polymers respond differently to radiation, it is useful to quantify the response, namely in terms of crosslinking and chain scission. A parameter called the G-value is frequently used for this purpose. It represents the chemical yield of crosslinks, scissions and double bonds, etc. For the crosslinked polymer, the crosslinking density increases with increasing the radiation dose, this is reflected by the swelling degree of the polymer while being immersed in a compatible solvent. If crosslinking predominates, the crosslinking density increases and the extent of swelling decreases. If chain scission predominates, the opposite occurs. A further detailed discussion of these aspects is presented throughout this chapter.

Keywords: gamma irradiation, polymerization, water radiolysis, radical

1. Introduction

Since the last quarter of the twentieth century, a growing interest in polymerization induced by ionizing radiation has been observed. Throughout this period, researchers have exposed polymeric materials to ionizing radiation and reported the occurring of crosslinking and other useful effects, e.g., miscibility of blend polymer [1–3] as well as many different applications in the polymer radiation chemistry field [4, 5]. Ionizing radiation is a promising technology for preparation of cost-effective, efficient, safe, and high-quality polymers. The radiation techniques applied for polymerization processes could be carried out using many different types of radiation sources. Ionizing radiations (gamma rays, X-rays, accelerated electrons, and ions beam) initiate polymerization reactions by the formation of very reactive intermediates (energetic radicals). Energetic radicals resulting from ionizing radiation hit molecules causing electrons ejection from a particular orbit, then the atom becomes ionized [6]. As soon as their orbits begin to be ready to overlapping forming new bonds, these electrons may be much more detached by other molecules [7, 8]. This phenomenon occurs when the transferred energy is higher than that of particular orbital electrons. When the energy is not high enough for ionization, electrons are gaining energy and transfer to an upper energy level, resulting in excitation. The ionizing potential for most molecules ranges from 7 to 15 eV [9], while the energies of ionizing irradiators (gamma rays, X-rays, accelerated electrons, and ions beam) range from 1 to 100 MeV [10], so ionization is the predominating process. Energetic radicals with ionized electron excite another molecule upon calling energy transfer. The ionized species associated with the energetic radicals created by irradiation would induce various reactions. This power in a polymerization reaction is an eco-friendly process. Polymers are covalently crosslinked by ionizing radiation at room temperature. The degree of crosslinking is radiation dose-dependent besides other parameters, e.g., temperature and ambiance [11].

Radiation crosslinking is a result of the combination of two molecules with active radicals at different three routes (head-to-head, head-to-tail, and tail-to-tail). In an amorphous region, the radical's combination is predominated rather than in crystalline region [12]. Sometimes increase of crystallinity reveals polymer crosslinking. On the other hand, the increase of amorphousity reveals polymer degradation (chain scission) [13]. Whatever, crosslinking or degradation, depends on the polymer nature. Therefore, predomination of radiation crosslinking can be achieved by removing oxidation gas (O_2) and replacing it by an inert gas (N_2). The percentage of dissolved oxygen could control the degree of polymerization and/or degradation of polymer chains. These processes compete through the breaking or formations of covalent bonds. The formations of new covalent bonds predominate in the case of vinyl monomer or synthetic polymer.

Water radiolysis produces neutral radicals $H^{\bullet}$ and $^{\bullet}OH$ with hydrated electrons e_{hy} and molecular products H_2 and H_2O_2. This means that the radiolysis of water gives several radiation-chemical reactions that have been studied in aqueous solutions, e.g., radiolytic reduction reaction (RRR), radiolytic oxidation reaction (ROR), and radiolytic polymerization reaction (RPR). This chapter discusses three kinds of RPR, namely; radical, cationic, and anionic polymerization.

In order to approach polymerization induced by ionizing radiation in high yield, the selected solvent should be ready for radiation radiolysis and of a high order of dissolving for the desired monomers. However, when switching to poor solvents, this might lead directly to rather broad distributions of polymer molecular weight. Non-polar solvents are ion-pairing or clustering that could interact, by radiolysis, with polymer backbone living a part of solvent molecules attached to polymer backbone "closely associated with the chain end," this is the so-called "living polymerization." Consequently, radiolytic solvents could act as catalysts which govern the kind of polymerization mechanism to be either cationic or anionic when solvents act as Lewis acid or Lewis base, respectively.

Step-growth radical polymerization induced by gamma irradiation occurs in three main stages. The first stage of the reaction is "initiated" by ionizing radiation, in which a radical (molecule with an active center) is formed, typically by radiolysis of an aqueous solvent (mainly water) or by braking C–H bond in polymer chains. The second stage is "propagation" which begins by attracting an active center that is formed in one monomer molecule with another monomer molecule. The final stage is "termination," in which the number of radical creation is terminated or the monomer molecules are consumed and mostly all monomers form polymer chains. Furthermore, the degree of polymerization depends on the monomer concentration and manipulation of the radiation dose.

Radiolysis of water is the ionization of water molecules by the absorption of ionizing radiation energy. The ionizing radiation absorbed by the water molecules is the pulse radiolysis technique [14]. The pulse radiolysis of water in high energy electromagnetic wave (ionizing radiation) gives variety of reactive species such as excited molecules, ions, and radicals [15]. It should be observed that radicals are formed by a wide variety of water radiolysis reactions, these species are chemical transformation and physical generation. The physical stage of the primary interaction of ionizing radiation with water molecules result in excitation and ionization as expressed by Eqs. (1) and (2):

$$H_2O \longrightarrow H_2O^* \text{ (Excitation)} \tag{1}$$

$$H_2O^* \longrightarrow H^\bullet + {}^\bullet OH \text{ (Ionization)} \tag{2}$$

This chapter focuses attention on the radiolysis of water as a source of radicals which are responsible for polymerization reactions. ESR spectroscopy is the best detection method for studying radical intermediates in radiolysis [16].

2. Photo-initiated induced by UV irradiation

2.1. Radical polymerization

Photopolymerization induced by UV irradiation is presented as a green technology characterized by low cost electrical power input and low energy requirements. Low temperature, low noise operation, and no volatile gases are generated. Photopolymerization is performed by water photolysis, practically the excited water molecule was generated $(H_2O)^*$ through the UV radiation ($\lambda < 240$ nm) absorption process (Eq. (3)). The water molecules in the excited state are mainly located on the water surface that exposed to UV, and with the course of time, energy will transfer to bulk molecules of water to produce (reactive species) radicals ($^\bullet H$ and $^\bullet OH$). The predominant photolysis reaction of UV in water molecules is homolytic bond scission (Eq. (4)). The excited water molecules $(H_2O)^*$ can undergo further photolysis producing reduction/oxidation agents (H^+ and ^-OH) by intra-molecular electron transfer (as summarized in Eq. (5)). The following equations are the most common water photolysis impact of UV absorption:

$$H_2O \longrightarrow H_2O^* \quad (3)$$

$$H_2O^* \longrightarrow H^\bullet + {}^\bullet OH \quad (4)$$

$$H_2O^* \longrightarrow H^+ + {}^-OH \quad (5)$$

In dissolving oxygen, the energy transfer processes to oxygen molecules forms dioxygen radical anion (superoxide radical) is possible according to Eq. (6). In case of complete energy transfer, relatively long-lived water molecules are produced in excited state, as the so-called the triplet state.

$$H_2O^* + O_2 \longrightarrow H_2O^{+\bullet} + {}^{\bullet -}O_2 \quad (6)$$

All the above reactions may be recombined with each other's in unlimited hypothesis, as expressed in Eqs. (7) and (8). This recombination gives excited $(H_2O)^*$ water molecules

$$H^\bullet + O_2 \longrightarrow HO^\bullet{}_2 \quad (7)$$

$$HO_2 + {}^\bullet OH \longrightarrow \boxed{{}^-OH + H^+} + {}^-O_2 \quad (8)$$

$$\boxed{{}^-OH + H^+} \downarrow H_2O^*$$

The hydroxyl radical ($^\bullet OH$) has a limited mobility and mainly abstract hydrogen atoms (H) for forming H_2O. On the other hand, the number of reduction/oxidation agents (H^+ and ^-OH) increases as the addition of oxygen to the water solutions increases. Obviously, at low concentration of dissolved oxygen, the crosslinked polymer is obtained. Therefore, under condition of sufficient dissolved oxygen, photo-oxidative degradation process may predominate. This is beneficial for water treatment and elimination of organic compounds pollutants; but sometimes, a part of the dissolved organic pollutants was found to precipitate, probably as a result of functional groups condensation (acetic acid, oxalic acid, n-butanoic acid, and malonic acid) which were the most abundant degradation products detected. The photo-oxidative degradation process C–C yields a macroradical $C^\bullet$ $C^\bullet$ by bond session. This macroradical molecule may be recombined in the absence of oxygen (polymerization). In the presence of oxygen, the peroxy radicals $C–O–O^\bullet$ may be formed (photo-oxidative degradation) [17]. Particularly, the photons that cause degradation are found in all electromagnetic radiations and can be enhanced at elevated temperatures. So, in polymerization reaction, the concentration of oxygen should be controlled in order to improve the polymerization induced by UV irradiation.

The role of the polymerization reaction initiation induced by UV irradiation "photoinitiator" is delayed with converting the photon energy into a chemical energy in the form of a "reactive center." When a water molecule absorbed photon in UV range in the absence of oxygen, it becomes excited and forms ($^{\bullet}$OH and $^{\bullet}$H) species with a high quantum yield [18]. The photoinitiator should exhibit an energetic wavelength, and undergo a rapid water photolysis to generate the initiating ($^{\bullet}$OH and $^{\bullet}$H). Subsequent polymerization goes on when reacting with a monomer molecule to start monomers combination and build 3D polymer network [19]. In this process, the polymers chains are crosslinked by forming covalent bonds cross chains [20].

The radical polymerization process is outlined in **Figures 1** and **2**, water excitation by absorbing UV leads to formation of free radicals as a first stage. In this typical step, the energetic radicals transfer to monomer molecules that undergo unimolecular bond formation upon irradiation. The polymerization process is complete with integration of monomer molecules. **Figure 2** shows water ionization by the UV rays that are capable to generate an active center (energetic radicals). These radicals pass energy to a monomer molecule that keeps the polymerization reaction going on. A "reactive" photoinitiator reacts with a monomer molecule to start a polymer chain formation, and the polymerization process begins with transferring the energetic radicals to monomer molecules. The crosslinked density is high in the deeper level rather than the surface level depending on oxygen concentration. While in free oxygen system, the crosslinked density is expected to be higher in the surface level rather than in the deeper level due to further yield of radicals on the side exposed to UV radiation.

2.2. Cationic polymerization

Besides photo-initiated radical polymerization that has been outlined, there are other two photo-polymerization processes; cationic and anionic polymerizations will be discussed in this section. Additives sensitive to UV radiation were required to promote cationic and anionic polymerization. In **Figures 3** and **4**, the common ultraviolet (UV) initiation agents are present

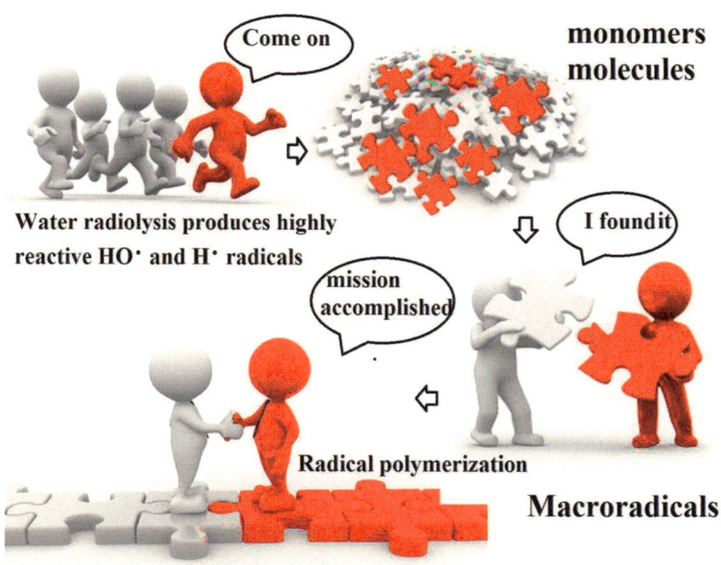

Figure 1. Imagination of the setup of radical polymerization.

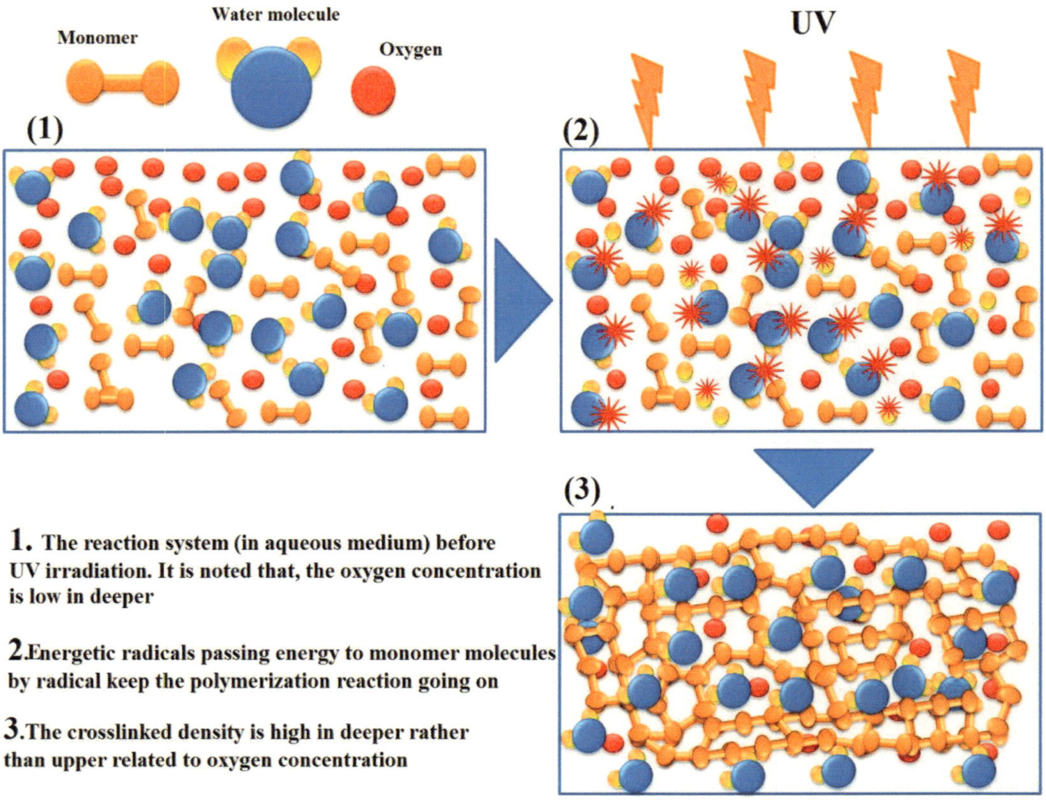

Figure 2. Proposed radical photo-polymerization induced by UV radiation.

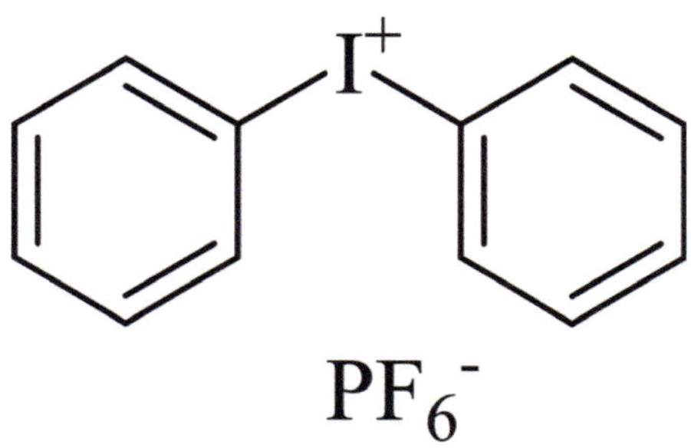

Figure 3. DPI diphenyliodonium hexafluorophosphate.

in two forms of onium salt: diphenyliodonium DPI (Ph_2I^+) and triphenylsulfonium DPS (Ph_3S^+) used as cationic photoinitiator under UV radiation [21–25]. Those are able to generate both protonic acids (carbonium ions) and free radicals (**Figure 5**).

This polymerization process is also called free-radical promoted cationic polymerization. This process undergoes more rapid polymerization, for example, in the curing process of epoxy resins, the ring opens and initiates both radical and cationic polymerization mechanism. Onium salt is sometimes called the curing agent, especially, for epoxy resin curing. **Figure 4**

http://dx.doi.org/10.5772/intechopen.73234

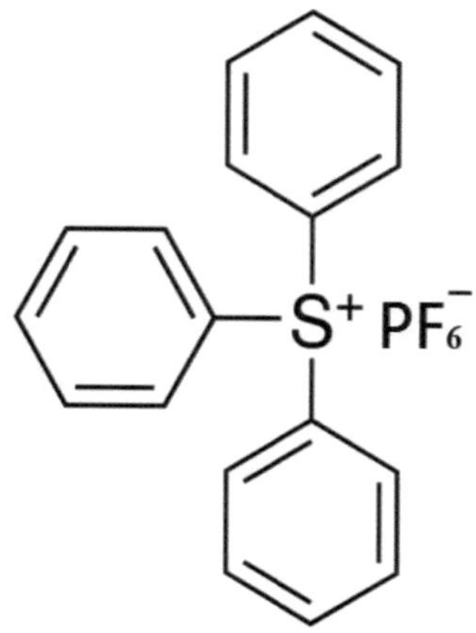

Figure 4. DPS triphenylsulfonium hexafluorophosphate.

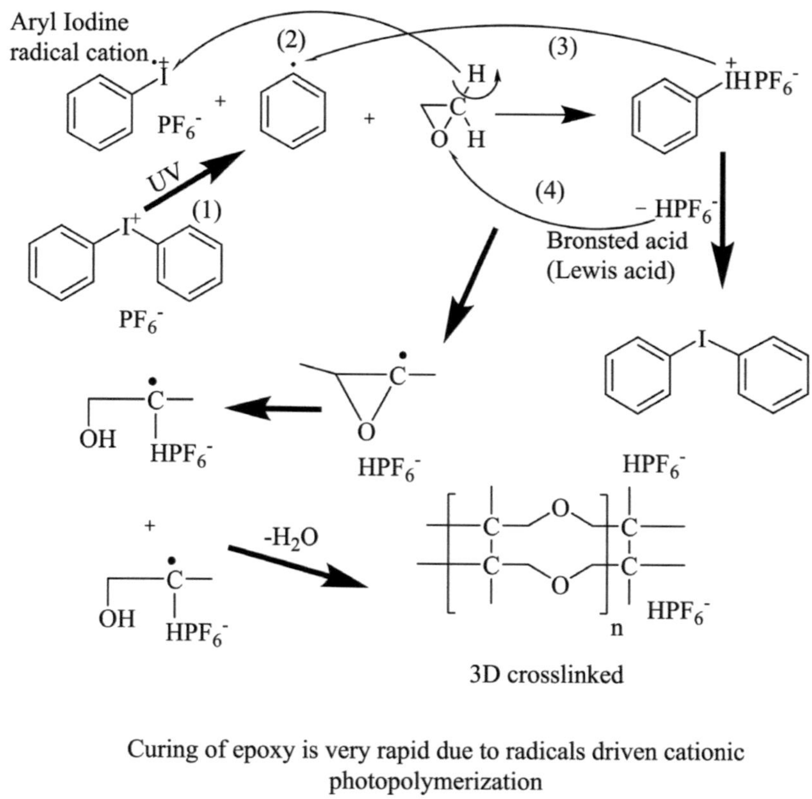

Figure 5. A proposed mechanism of radical cationic photopolymerization induced by UV radiation of epoxy resin.

shows the generation of a Bronsted acid (Lewis acid) when DPI and DPS exposed to UV radiation [26–28].

As depicted in **Figure 5**, an aryl iodine radical cation and aryl radical are formed by the photolysis of DPI salt (1). Both of these are highly reactive species which separate protons from the monomer (2) generating bronsted acid $(HPF_6)^-$ and formatting diaryl iodine (3). The polymerization reaction rapidly involves by protonation of the epoxide oxygen with the very powerful acid (4). It is obvious that the cationic polymerization of epoxies is highly quantitative and very rapid, with low activation energy [29]. For clarity, the concurrent radical and cationic reactions in hybrid systems have dramatically changed the rate of photo-polymerization [30].

Figure 5 shows a proposed mechanism of radical cationic photo-polymerization induced by UV radiation of epoxy resin.

2.3. Anionic polymerization

Free cationic and radical photopolymerizations have been extensively applied in this context, whereas photo-induced anionic polymerization was limited in scope and application [31]. Kahveci et al. used a combination of cationic photo-polymerization and anionic opening ring polymerization mechanisms to confirm the amphiphilic graft copolymers of poly(ethylene oxide)-*g*-poly (isobutyl vinyl ether) (PEO-*g*-PIBVE) [32]. First, poly(ethylene oxide-co-ethoxyl vinyl glycidyl ether) was synthesized by anionic ring-opening co-polymerization of corresponding monomers using radiation for reaction initiation. Random anionic ring-opening co-polymerization of a mixture of the gaseous EO (b.p. 11°C) and the respective co-monomer in an appropriate solvent can be initiated by an alkali-metal alkoxide [33]. The vinyl ether moieties were then cationic photo polymerized using diphenyliodonium iodide photoinitiator. It is clear that photoinitiator was only used for cationic polymerization, while anionic polymerization takes place spontaneously by ring opening in mild conditions. Further investigation is needed to cover unexplored areas in the field of photo-induced polymerization [34]. Radicals promote cationic photo-polymerization, since it has been demonstrated to be a very effective method for the preparation of well-defined high molecular weight epoxy resin [35].

3. Electron beam

In order to complete the image of radiation-induced polymerization, another radiolysis system should be taken into consideration is electron beam. Electrons are generated in a linear beam form causing a powerful and fast water radiolysis that produce hydrated electrons (e_{hy}). Most solvents are of a continuum for some polarity properties. The hydrated electron (e_{hy}) is very reductive chemical species, once the solvated electron hits water molecules, water becomes ionized [36]. This depends on the time consumed for interaction and the average distance between the solvated electron (e_{hy}) and the water ions.

The emitting powerful electron (more than 1 MeV electron) from electron accelerator hits water molecule and then scatters. The energy of electron is lost as a result of bordering a pathway for secondary electrons. Secondary electrons keep their own powerful energy, causing water ionization and forming copious quantities of variety of reactive species (e.g., H_2 H_2O^+, H_2O_2,

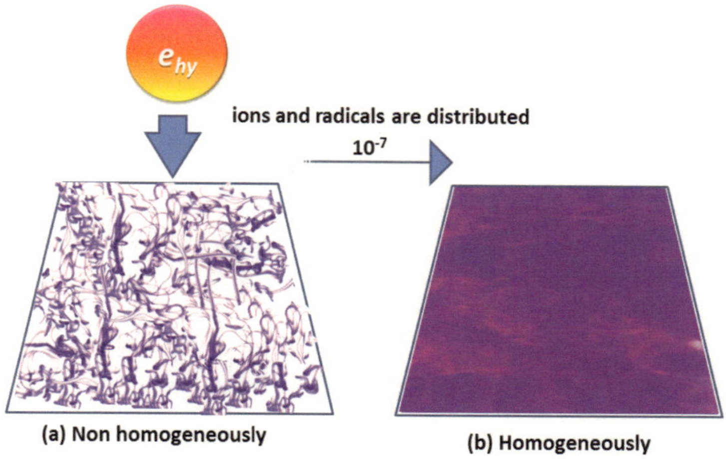

Figure 6. Inhomogeneous distribution of ionizing events (e_{hy}) and its diffusion with time (~10^{-7}) and reactions reach the formation homogenous distribution.

OH^-, H_3O^+ and $^{\bullet}OH$, etc.) [37]. Particularly, the chemical reactions (radiolysis) occur when the reactive species of ions and radicals are distributed nonhomogeneously in "clusters" of excitations/ionization (i.e., spurs), and homogeneously after the reactive species have diffused and thermalized [38]. Spurs were used to describe the high energy electron passed through liquids. It is well-known that the ionizing and exciting species are, in fact, nonhomogeneous and occur in clusters called "spurs." The electron is similar to a "stone" hit in the surface of the water pool, by the time the resulting "splashes" induced by "the stone hit" will expand and eventually overlap with their neighbors to result in a homogenous distribution (**Figure 6**). However, this simple picture is not true in some cases; the irradiation of nonaqueous systems does not yield as a clear picture concerning spurs as the case in the water.

4. X-ray

Risse and Fricke [39, 40] had studied the radiolysis of water by X-rays in a closed vessel. Bubbles of hydrogen and carbon dioxide gas resulting from irradiation are found [41]. X-ray photons are highly powerful causing radiolysis of water to create an ion pair $(H_2O)^+$ (e_{hy}). The amount of ionization species produced is affected by the amount of energetic photons absorbed. **Figure 7** shows that the positive and negative water molecules are unstable. These can break apart into smaller molecules. Free radicals can be formed by this breakup. Several chemical reactions might occur (oxidation reduction) by (OH^-, H^+, and e_{hy}) beside polymerization takes place by (•OH and •H and e_{hy}) [42]. Also, the radiolysis products may recombine individually, forming gases such as H_2 and H_2O_2 [43], and hence the bubbles are formed due to an evolving mixture of H_2 and O_2 gases. H_2 present in the solution, or the "e_{hy}," may cause reduction reactions.

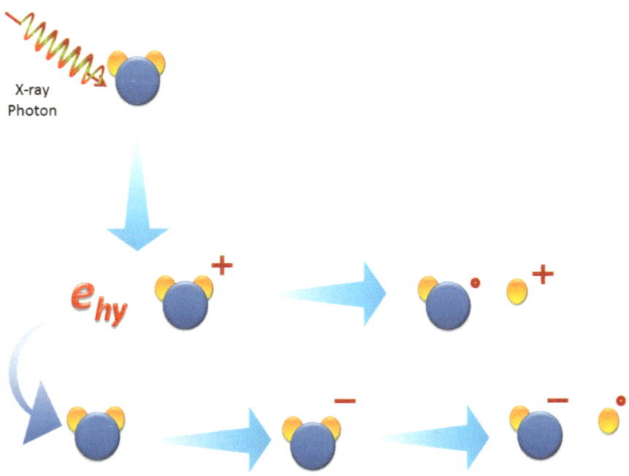

Figure 7. In radiolysis, bulk water is ionized to form an $(H_2O^+)_{aq}/(e^-)_{aq}$ pair. The $(H_2O^+)_{aq}$ quickly converts to ($^{\bullet}$OH) and $^+$H to become $(H_2)_{aq}$. And, the hydrated electron (e_{hy}) hits another water molecule to become $(H_2O^-)_{aq}$, which converts to (OH^-) and $^{\bullet}$H to become $(H_2)_{aq}$.

5. Gamma radiation

Radioisotopes, e.g., (Co-60 $h_{1/2}$ = 5.3 years, Cs-137 $h_{1/2}$ = 30 years, and Ir-192 $h_{1/2}$ = 74 days) are considered as the sources of gamma rays. The radiation chemistry of polymers dissolved in water is the chemistry of reactions with $^{\bullet}$OH, $^{\bullet}$H, and e_{hy}. The effect of ionizing radiation on water was first studied in 1913 by Duane and Scheuer [44] and the results were confirmed by Lanning and Lind after 25 years at 1938 [45]. Several chemical reactions occur in irradiated systems both indirect and direct reactions which take place homogeneously. The latter reactions are studied in details for the case of water radiolysis by γ-irradiation. The ultra-pure water irradiated by γ rays produces 11 reactive species resulted from 20 reactions, while in acidic water it produces 7 reactive species resulted from 10 reactions [46].

In ultra-pure irradiated water, there are no impurities that can scavenge radicals and prevent the tracks of radiolysis to expand. The increase of radical concentrations leads to a homogeneous radical-radical combination (H_2, H_2O_2, and O_2). The excess radicals also react with the products of (H_2, H_2O_2, and O_2) in homogeneous reactions. All of these phenomena must be considered in polymerization reaction induced by gamma irradiation. Two cases have been established in the polymerization induced by ionizing radiation, namely, crosslinking and degradation. In the polymerization system that starts with crosslinkable monomer or polymer, the system will consume radicals ($^{\bullet}$H and $^{\bullet}$OH) in polymerization reaction and limits build-up of (H_2, H_2O_2, and O_2). While in the system that tends to degradable polymer such as natural polymer, the yield of radicals' duplicates and the breaks of C–C bonds increase and tend to enhance the polymer degradation by formation of a net of oxidative spices such as H_2O_2, O_2, and O_3.

It is interesting to investigate the proposed mechanism of the polymerization reaction induced by gamma irradiation in an aqueous system. This section, presents the steps of common three polymerization mechanisms (radical, cationic, and anionic). It should be observed that polymerization reaction randomly forms covalent bonds cross the monomer and short chains in space.

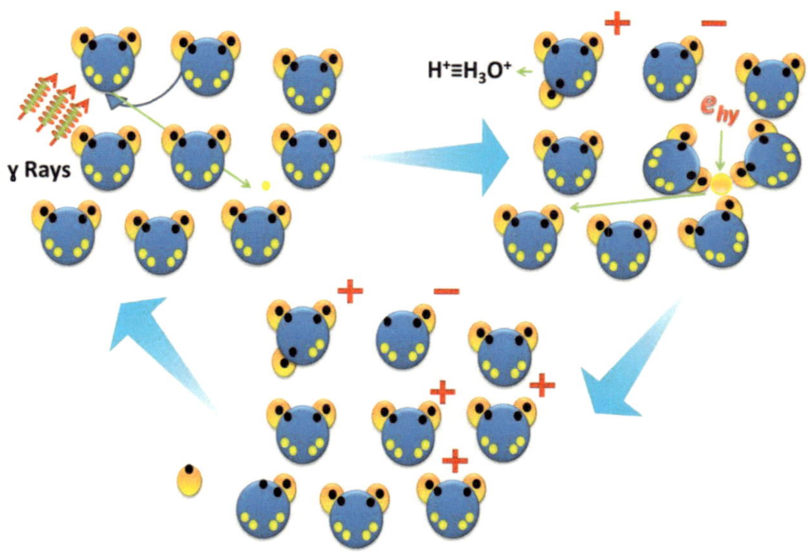

Figure 8. The close system of water radiolysis induced by gamma rays.

A background of water radiolysis is expected to fluctuate a detection of the majority of primary products of water ions formation (H_3O^+, HO^-, H^+ ions, and e_{hy}). A track of initial species formation is shown in **Figure 8**.

Ionization of H_2O occurs as: $H^+ \equiv H_3O^+$ and OH^-. The question here is how this really happened in ultra-pure water solution that contains water molecules? The oxygen molecule with two lone pair of electrons and two hydrogen atoms react with another water molecule under the impact of high power electromagnetic wave (ionizing radiation). The H^+ atom interacts with H_2O to form H_3O^+ and ^-OH. This structure (H_3O^+) is to great extent, similar to H^+. The production of hydrated electron (e_{hy}) was detected surrounded with opposite charge of water molecules. This (e_{hy}) hits one water molecule to produce two radicals ($^\bullet H$ and $^\bullet OH$) leaves the surrounded water molecules as (H_2O^+). After finishing the exposure to ionizing radiation, the water molecule returns to the initial state. The equations of radical production by water ionization by (e_{hy}) are as follows:

$$H_3O^+ + e_{hy} \longrightarrow 2H^\bullet + {}^\bullet OH \tag{9}$$

$$H_2O^+ + e_{hy} \longrightarrow H^\bullet + {}^\bullet OH \tag{10}$$

$$H^+ + e_{hy} + H_2O \longrightarrow 2H^\bullet + {}^\bullet OH \tag{11}$$

The polymerization reactions depend on several parameters: (1) chemical structure of the polymer (e.g., monomer containing aromatic rings has a much greater resistance to radiation crosslinking rather than the aliphatic monomer), (2) pH and temperature of the polymerization system, (3) kind and composition of the appropriate solvent, and (4) exposure time. The control of all of these parameters facilitates the modification of polymers by radiation techniques. Polymerization reactions are generally classified as follows: (1) predominantly undergoing crosslinking that gives hydrogel when the polymer has hydrophilic character, (2) grafted film, (3) dimerization, and (4) oligomerization.

5.1. Radical polymerization induced by gamma irradiation

Radical initiation is an easy and a suitable method for the synthesis of hydrogels. When vinyl monomers are exposed to ionizing radiation in aqueous media (e.g., water), the radiolysis of water leads to formation of radicals that are responsible for initiating the polymerization reactions. In the propagation steps, the rapid recombination of two monomers takes place [47–49]. **Figure 9** shows the proposed mechanism of ionizing radiation-induced hydrogel formation. An example for this is hydrophilic crosslinked hydrogel. When the irradiated polymer and/or monomer is dissolved in aqueous solution, reactive radicals on $C^{\bullet}$ atom are formed [50]. The main reactive species generated in water by irradiation are hydroxyl radicals $^{\bullet}OH$, hydrated electrons (e_{hy}), and hydrogen radicals $^{\bullet}H$. The hydrated electron (e_{hy}) reacts with water molecules to produce $^{\bullet}OH$ and $^{\bullet}H$ radicals (**Figure 1**). The direct action of e_{hy} is very weak [51]. However, two $^{\bullet}H$ radicals combine fast to form H_2 gas. **Figure 9** demonstrates the proposed mechanism of PEO (poly ethylene oxide) radical polymerization. The main reactive species ($^{\bullet}OH$ radical) are responsible for the polymerization process. In general, the number of radicals generated in water is higher than that in polymer and/or monomer itself. Therefore, the description in **Figure 9** below will only refer to the indirect mechanism. Hydroxyl ($^{\bullet}OH$)/abstract hydrogen atoms from C–H in polymer molecules leads to form macroradical ($C^{\bullet}$). Thus, macroradical (polymer radicals) causes 3D crosslinked polymer. The rate constant of the polymerization reactions is proportional to the number of radical's production. Details are shown in **Figure 9** for only polyethylene oxide, all the hydrogen atoms are equivalent in the polymer molecule, so that the only one type of macroradicals $C^{\bullet}$ can be

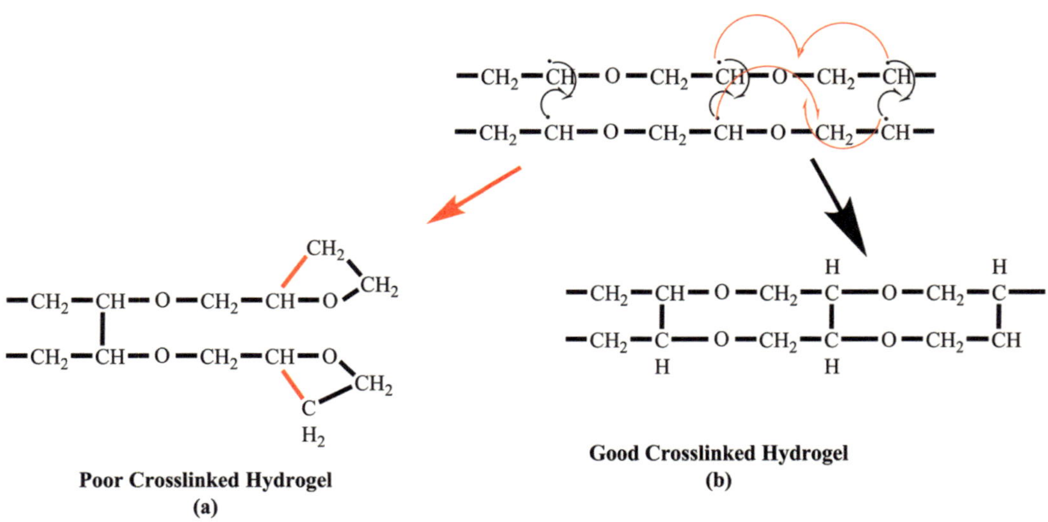

Figure 9. Proposed macroradicals formation induced by ionizing irradiation of water-soluble PEO in aqueous solution.

http://dx.doi.org/10.5772/intechopen.73234

formed by •OH radicals abstraction of hydrogen. While in non-equivalent H atoms positions in polymer backbone, hydroxyl radical abstract hydrogen atoms from various attitudes (randomly) lead to the formation of different kinds of macroradicals of different polymer structure.

As seen in **Figure 9**, macroradicals may undergo *intra*-molecular crosslinking giving PEO hydrogel network with randomly covalent bonds beside H bonds. The inter-molecular crosslinking is the most important reaction by "recombination" of macroradicals localized on two different chains, at this condition, the yield of crosslinked hydrogel will increase [52] (**Figure 9b**). Furthermore, inter-molecular "disproportionation" involves either linked radical or scission in the same chain. This process does not lead to hydrogel formation [53] (**Figure 9a**). There is a competition between *inter*- and *intra*-molecular crosslinking fully recognized in using ionizing radiation. Energetic photon could increase the yield of inter-molecular crosslinks (Gx), defined as the number of crosslinks formed in the system upon absorption of 1 J of ionizing radiation energy [54]. So the formation of hydrogel network depends on the energy of interchain bonds formation. The radicals combine by various ways that can be estimated in a number of routes. **Figure 10** presents various proposed routes of crosslinked formation of PVA. As illustrated in the figure, the hydrogen bond formation between chains increases the chance of polymerization mechanism to undergo intra-molecular "recombination" crosslinking rather than inter-molecular "disproportionation." However, in the case of oxygen gas, "disproportionation" reactions will be predominantly leading to the formation of carbonyl groups. In the presence of oxygen, the PVA radicals are converted into the corresponding (HOO•) peroxyl radicals leading to strand breakage [55]. It is well-known that the hydrogel formation is of high yields in oxygen-free systems.

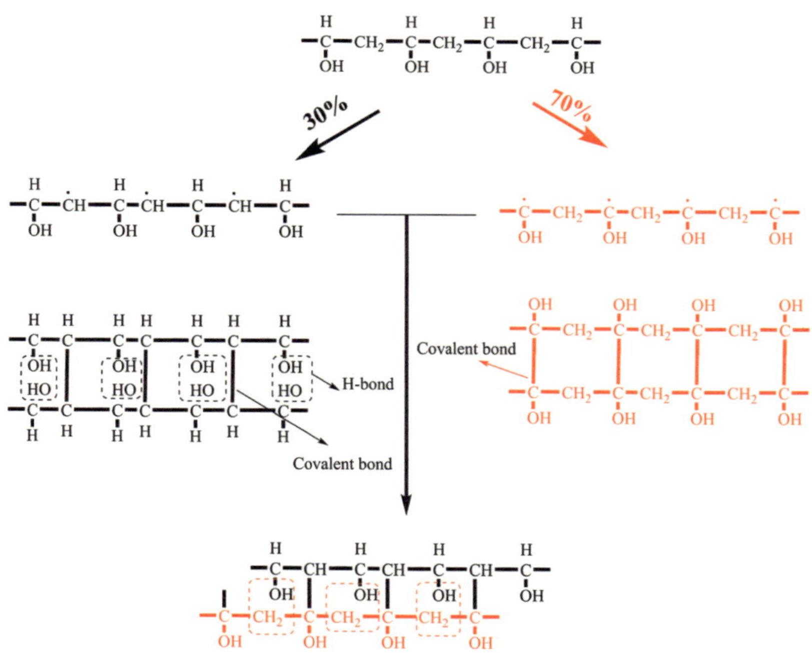

Figure 10. Proposed macroradicals formation induced by ionizing irradiation of water-soluble PVA in aqueous solution.

5.2. Ionic polymerization: anionic and cationic polymerization

Cationic and anionic polymerization induced by ionizing radiation are more complicated than radical polymerization mechanism, where cationic and anionic polymerization require some conditions and additives that act as Lewis acid and base, respectively. Gamma irradiation is very useful to obtain these additives by solvent radiolysis. Before formulating the main steps of the ionic polymerization mechanisms in radiation initiation, this section will consider the main concepts forcing an ionizing radiation-induced ionic polymerization. First, the nature of chemical structure of monomers, e.g., for cationic mechanism is vinyl ethers, isobutylene, β-pinene, and cyclopentadiene, and for anionic mechanism is vinylidene cyanide and nitro-ethylene. Second, it is necessary to use the radiation tool for formation and stabilization of **C^+** (***Carbonium***) in cationic and **C^-** (***Carbanion***) in anionic mechanisms. Third, choosing appropriate halogenated solvents for cationic polymerization, e.g., methylene chloride, ethyl chloride, tetrafluoromethane, difluorodichloromethane, perchloro ethylene, etc. These solvents generally undergo radiolysis to yield Lewis acid (electron acceptor) from a Lewis base (polymer) and to form "***adduct***." Solvent for anionic polymerization, e.g., triethylamine, dimethyl formamide, tetrahydrofuran, isopropylamine, ethylpropyl ketone, acetone, pyridine, etc. These solvents that generally undergo radiolysis to yield Lewis base (electron donating) are suitable for anionic mechanism.

Some compounds such as solid additives, catalyze the radiation-induced cationic polymerization, e.g., calcium, zinc, silica gel, and magnesium oxides [56, 57]. Even the monomer can be polymerized by combining the two mechanisms (radical/anionic or radical/cationic), then by changing the conditions (temperature and solvent).

5.2.1. Cationic polymerization

The rate of polymerization by cationic mechanism depends on the stability of carbonium ions in the macromolecules. This can be achieved if the center of the charged ion is strongly solvated by opposite charge and the distance between solvated ions is far. For the carbonium ion to be sufficiently stable and have enough lifetimes, it is preferred to choose a monomer that has a sufficient nucleophilic character. Examples of active monomers that can undergo cationic polymerization by gamma radiation are styrene monomer. In this monomer, the combination of a carbonium with an electron donating (Lewis acid) leads to

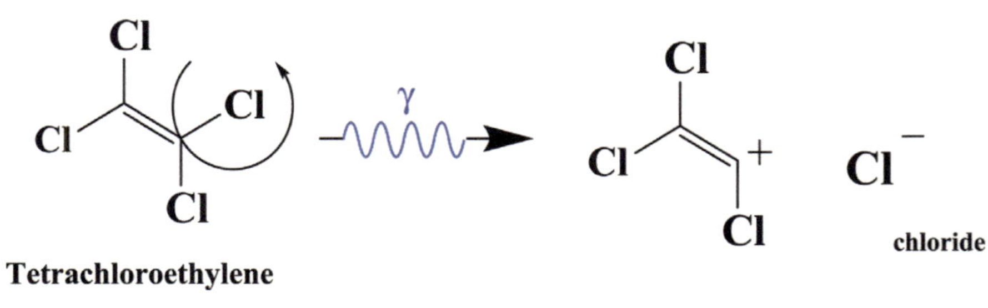

Figure 11. The essential step is the formation of Lewis acid by solvent radiolysis.

cationic polymerization reaction. **Figure 11** shows the essential step of the formation of Lewis acid by solvent radiolysis. For that tetrachloroethylene, a chlorinated hydrocarbon solvent, could be used as effective reaction media to promote the cationic polymerization of styrene [58].

The gamma radiolysis (γ-rays) of tetrachloroethylene liberates: (i) chloride ions (Cl^-), (ii) carbonium species (C^+) that act as Lewis acid species [59], and/or (iii) combined cation radical species ($C^{+\bullet}$) [60–62], which can induce PS linear chain recombination (crosslinking). The proposed mechanism, illustrated in **Figure 12**, of radiation-induced cationic polymerization of linear polystyrene in tetrachloroethylene gave rather good reproducibility and shows a good agreement with the results of Ueno et al. [63]. Also, in the polymerization of styrene in methylene chloride, Tsuda [64] has suggested that hydrogen chloride produced by the irradiation of chlorinated hydrocarbons causing initiation of cationic polymerization of styrene. It is

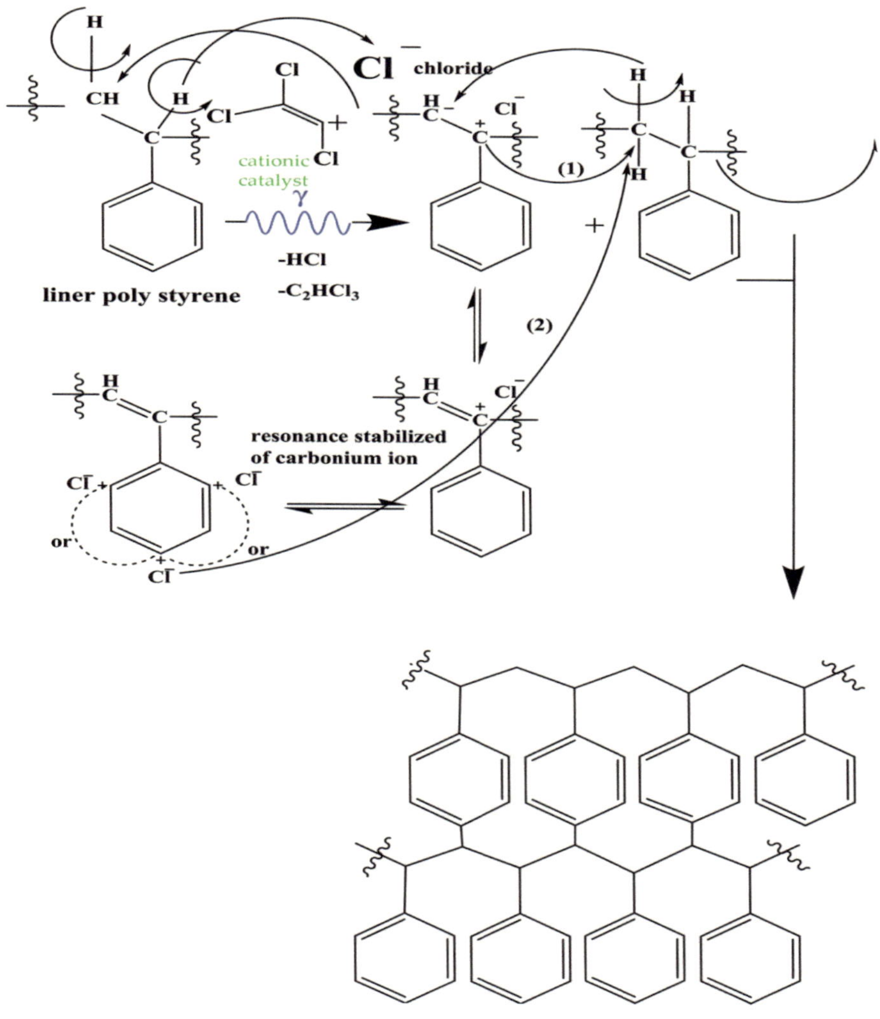

Figure 12. The proposed mechanism of styrene radical polymerization induced by gamma irradiation.

well-known that in chlorinated solvent, the polymerization of styrene proceeds well by cationic polymerization. This proposed mechanism is in a good agreement with studies carried out by Sigwalt and Moreau [65]. As shown in **Figure 12**, the initiation starts by the formation of Cl^- due to solvent radiolysis which is released as HCl giving a carbon cation on styrene unites. Propagation step is formed by head-to-tail addition in carbocation polymerization. β-Hydrogen atoms of the propagating carbonium cation are quite acidic, owing to the delocalization of the positive charge [66] belonging to the phenyl ring. The transfer of positive charge in the phenyl ring gives a chance of addition on the phenyl ring causing the formation of 3D crosslinked polymer. Still cationic polymerization may be one of the most challenging areas of future research [67]. The final structure as seen in an ideal shape (head-to-tail), sometimes, the propagating chain may rearrange randomly.

6. Advantages of ionizing radiation-induced polymerization

The advantages are:

1. Easy process of polymerizing a wide range of monomers which are resistant to polymerization reaction by conventional methods [68].

2. Very safe, in case of polymerization involving hazardous compounds (e.g., Organo chlorine) that can be decomposed by radiation [69, 70].

3. Polymerization reaction in the solid state could be carried out [71, 72].

4. Increased purity of polymer obtained without by product; while in conventional methods, it needs complicated purification steps [73, 74].

7. Concluding remarks

This chapter presents a very brief introduction of the interactions between high energy photons with water molecules, and the subsequent physico-chemical stage of water radiolysis based on previous literatures. Some proposed mechanisms in radiation polymerization were also outlined in this chapter. The main reactive species ((e_{hy}) hydrated electron, $^{\bullet}OH$ radicals, and $H^{\bullet}$ radicals) generated from water radiolysis are presented. The ionized species of water molecules involving them are ultrafast, occurring on femtosecond. The reason to focus on water radiolysis is the reactivity of radicals and hydrated electron, which are the initiators of polymerization reactions. Irradiation can be very useful in the processing of polymer blends, which often undergo physical phase separation of components due to incompatibility of materials. In a number of blend systems, irradiation has been used to induce crosslinking of one or more of the components and/or formation of crosslinking between the different phases, resulting in improvement of physical properties.

8. Image gently advocates on the safety of radiation uses

Author details

Mohamed Mohamady Ghobashy

Address all correspondence to: mohamed_ghobashy@yahoo.com; mohamed.ghobashy@eaea.org.eg

Radiation Research of Polymer Department, National Center for Radiation Research and Technology (NCRRT), Atomic Energy Authority, Nasr City, Cairo, Egypt

References

[1] Ghobashy MM, Khozemey E. Sulfonated gamma-irradiated blend poly(styrene/ethylene-vinyl acetate) membrane and their electrical properties. Advances in Polymer Technology. 2016. DOI: 10.1002/adv.21781

[2] Ghobashy MM. Effect of sulfonated groups on the proton and methanol transport behavior of irradiated PS/PEVA membrane. International Journal of Plastics Technology. 2017; **21**(1):130-143. DOI: 10.1007/s12588-017-9176-5

[3] Ghobashy MM, Abdel Reheem AM, Mazied NA. Ion etching induced surface patterns of blend polymer (poly ethylene glycol–poly methyl methacrylate) irradiated with gamma rays. International Polymer Processing. 2017;**32**(2):174-182

[4] Ghobashy MM. *In-situ* core-shell polymerization of magnetic polymer nanocomposite (PAAc/Fe_3O_4) particles via gamma radiation. Nanocomposites. 2017;**3**(1):42-46. DOI: 10.1080/20550324.2017.1316600

[5] Ghobashy MM, Khafaga MR. Chemical modification of nano polyacrylonitrile prepared by emulsion polymerization induced by gamma radiation and their use for removal of some metal ions. Journal of Polymers and the Environment. 2017;**25**:343. DOI: 10.1007/s10924-016-0805-4

[6] Chapiro A. Radiation chemistry of polymers. Radiation Research Supplement. 1964;**4**:179-191

[7] Böhler E, Warneke J, Swiderek P. Control of chemical reactions and synthesis by low-energy electrons. Chemical Society Reviews. 2013;**42**(24):9219-9231

[8] Charlesby A. Atomic Radiation and Polymers: International Series of Monographs on Radiation Effects in Materials. Shrivenham, UK: Elsevier; 2016

[9] Gross JH. Principles of ionization and ion dissociation. In: Mass Spectrometry. Berlin, Heidelberg: Springer International Publishing; 2017. pp. 29-84

[10] Chmielewski AG, Haji-Saeid M, Ahmed S. Progress in radiation processing of polymers. Nuclear Instruments and Methods in Physics Research Section B: Beam Interactions with Materials and Atoms. 2005;**236**(1):44-54

[11] da Silva A, Aparecida K. Sterilization by gamma irradiation. In: Gamma Radiation. Rijeka, Croatia: InTech; 2012

[12] Arutchelvi J, Sudhakar M, Arkatkar A, Doble M, Bhaduri S, Uppara PV. Biodegradation of polyethylene and polypropylene. Indian Journal of Biotechnology. 2008;**7**:9-22

[13] Czvikovszky T. Degradation effects in polymers. In: Advances in Radiation Chemistry of Polymers. Proceedings of a Technical Meeting held in Notre Dame, Indiana, USA; 2004. pp. 91-102

[14] Bensasson RV, Land EJ, Truscott TG. Flash Photolysis and Pulse Radiolysis: Contributions to the Chemistry of Biology and Medicine. Elsevier: Pergamon Press; 1983

[15] Slegers C. A Concise Introduction on Radiation Chemistry. Bruxelles: Chimie Nouvelle. 2006;**91**:17

[16] Pryor W, editor. Free Radicals in Biology. Vol. 6. USA: Elsevier; 2012

[17] Potts JE. Plastic environmentally degradation. In: The Kirk-Othmer Encyclopedia of Chemical Technology. Vol. 3. USA: John Wiley and Sons, Inc.; 1991

[18] Decker C, Elazouk B. Laser curing of photopolymers. In: Allen NS et al., editors. Current Trends in Polymer Photochemistry. New York: Ellis Horwood; 1995. p. P130

[19] Jacobs PF. Rapid Prototyping & Manufacturing, Fundamentals of Stereolithography. New York, NY: Society of Manufacturing Engineers; 1992

[20] Crivello JV, Dietliker K. Photoinitiators for free radical, cationic & anionic photopolymerisation. In: Bradley G, editor. Chemistry & Technology of UV & EB Formulation for Coatings, Inks & Paints. 2nd ed. Vol. III. Chichester & New York: John Wiley & Sons, Inc.; 1998. in association with SITA Technology Ltd., London, UK

[21] Crivello JV, Lam JHW. Diaryliodonium salts: A new class of photoinitiators for cationic polymerization. Macromolecules. 1977;**10**:1307-1315

[22] Crivello JV, Lam JHW. Photoinitiated cationic polymerization with triarylsulfonium salts. Journal of Polymer Science, Polymer Chemistry Edition. 1979;**17**:977-999

[23] Crivello JV, Aldersley MF. Supramolecular diaryliodonium salt-crown ether complexes as cationic photoinitiators. Journal of Polymer Science, Part A: Polymer Chemistry. 2013;**51**: 801-814

[24] Crivello JV, Juliano PC. Polyimidothioether–polysulfide block polymers. Journal of Polymer Science Part A: Polymer Chemistry. 1975;**13**(8):1819-1842

[25] Ashcroft WR. Curing agents for epoxy resins. In: Chemistry and Technology of Epoxy Resins. Dordrecht, The Netherlands: Springer Science+Business Media; 1993. pp. 37-71

[26] Sasaki Y. U.S. Patent No. 5,480,918. Washington, DC: U.S. Patent and Trademark Office; 1996

[27] Sasaki Y. U.S. Patent No. 5,389,700. Washington, DC: U.S. Patent and Trademark Office; 1995

[28] Felder L, Kirchmayr R, Bellus D. U.S. Patent No. 4,088,554. Washington, DC: U.S. Patent and Trademark Office; 1978

[29] Crivello JV, Liu S. Photoinitiated cationic polymerization of epoxy alcohol monomers. Journal of Polymer Science Part A: Polymer Chemistry. 2000;**38**(3):389-401

[30] Bassi GL. Formulation of UV-curable coatings—How to design specific properties. In: Fouassier JP, Rabek JF, editors. Radiation Curing in Polymer Science and Technology—Vol II: Photoinitiated Systems. London & New York: Elsevier Applied Science; 1993. p. P239

[31] Dadashi-Silab S, Doran S, Yagci Y. Photoinduced electron transfer reactions for macromolecular syntheses. Chemical Reviews. 2016;**116**(17):10212-10275

[32] Kahveci MU, Mangold C, Frey H, Yagci Y. Graft copolymers with complex polyether structures: Poly (ethylene oxide)-graft-poly (isobutyl vinyl ether) by combination of living anionic and photoinduced cationic graft polymerization. Macromolecular Chemistry and Physics. 2014;**215**(6):566-571

[33] Obermeier B, Wurm F, Mangold C, Frey H. Multifunctional poly(ethylene glycol)s. Angewandte Chemie International Edition. 2011;**50**(35):7988-7997

[34] Weiss P. Photo-induced polymerization. Pure and Applied Chemistry. 1967;**15**(3-4):587-600

[35] Atif M, Bongiovanni R, Yang J. Cationically UV-cured epoxy composites. Polymer Reviews. 2015;**55**(1):90-106

[36] Jay-Gerin JP, Ferradini C. Compilation of some physicochemical properties of solvated electrons in polar liquids. Journal de Chimie Physique. 1994;**91**:173-187

[37] Garrett BC, Colson SD, Dixon DA, Laufer AH, Ray D. Understanding the Role of Water on Electron-Initiated Processes and Radical Chemistry. United States: USDOE Office of Science (SC); 2003

[38] Mozumder A, Magee JL. Theory of radiation chemistry. VII. Structure and a reactions in low linear energy transfer tracks. The Journal of Chemical Physics. 1966;**45**:3332-3341

[39] Risse O. About the X-ray photolysis of hydroperoxide (in German). Zeitschrift für Physikalische Chemie. 1929;**140**(1):133-157

[40] Fricke H, Hart EJ, Smith HP. Chemical reactions of organic compounds with X-ray activated water. The Journal of Chemical Physics. 1938;**6**(5):229-240

[41] Gunther P, Holzapfel L. The X-ray sensitivity of liquid water and ice. Zeitschrift für Physikalische Chemie. 1939;**446**:374-382

[42] Mesu JG, Beale AM, De Groot FM, Weckhuysen BM. Observing the influence of X-rays on aqueous copper solutions by in situ combined video/XAFS/UV-Vis spectroscopy. In: AIP Conference Proceedings (Vol. 882, No. 1, pp. 818-820). UK: AIP; 2007, February

[43] Stein G, editor. Radiation chemistry of aqueous systems, the Weizmann science press of Israel, Jerusalem, (1968) B. Pastina, J. A. LaVerne, S. M. Pimblott. The Journal of Physical Chemistry. A. 1999;**103**:5841

[44] Duane W, Scheuer O. Decomposition of water by a-rays. Radium. 1913;**10**:33-46

[45] Lanning FC, Lind SC. Chemical action of alpha particles from radon on aqueous solutions. The Journal of Physical Chemistry. 1938;**42**(9):1229-1247

[46] Chatterjee A, Magee JL, Dey SK. The role of homogeneous reactions in the radiolysis of water. Radiation Research. 1983;**96**(1):1-19

[47] Charlesby A. Atomic Radiation and Polymers. Oxford: Pergamon Press; 1960

[48] de Rossi D, Kajiware K, Osada Y, Yamauchi A. Polymer Gels. Fundamentals and Applications. New York: Plenum Press; 1991

[49] Peppas NA, Mikos AG. Preparation methods and structure of hydrogels. Hydrogels in Medicine and Pharmacy. 1986;**1**:1-27

[50] von Sonntag C. The Chemical Basis of Radiation Biology. London: Taylor & Francis; 1987

[51] Rosiak JM, Ulanski P. Synthesis of hydrogels by irradiation of polymers in aqueous solution. Radiation Physics and Chemistry. 1999;**55**:139

[52] Alexander P, Charlesby A. Effect of X-rays and γ-rays on synthetic polymers in aqueous solution. Journal of Polymer Science. 1957;**23**:355

[53] Ulanski P, Bothe E, Hildenbrand K, Rosiak JM, von Sonntag C. Hydroxyl-radical-induced reactions of poly(acrylic acid); a pulse radiolysis, e.p.r. and product study. Part I. Deoxygenated aqueous solutions. Journal of the Chemical Society, Perkin Transactions II. 1996;**1996**:13

[54] von Sonntag C. The Chemical Basis of Radiation Biology. London: Taylor & Francis; 1987

[55] Ulanski P, Bothe E, Rosiak JM, von Sonntag C. OH-radical-induced crosslinking and strand breakage of poly(vinyl alcohol) in aqueous solution in the absence and presence of oxygen. A pulse radiolysis and product study. Macromolecular Chemistry and Physics. 1994;**195**:1443

[56] Crivello JV. UV and electron beam-induced cationic polymerization. Nuclear Instruments and Methods in Physics Research B. 1999;**151**:8-21

[57] Crivello JV. Advanced curing technologies using photo- and electron beam a induced cationic polymerization. Radiation Physics and Chemistry. 2002;**63**:21-27

[58] Chapiro A. Cationic polymerizations initiated by high energy radiation. Macromolecular Chemistry and Physics. 1974;**175**:1181-1197

[59] Sutherland JW, Spinks JWT. Radiolysis of Tetrachloroethylene. Canadian Journal of Chemistry. 1959;**37**:79-90

[60] Rajesh P, LaVerne JA, Pimblott SM. High dose radiolysis of aqueous solutions of chloromethanes: Importance in the storage of radioactive organic wastes. Journal of Nuclear Materials. 2007;**361**:10-17

[61] Popova AI, Sheinker AP, Abkin AD. Radiation polymerization of isobutylene in various solvents. Polymer Science U.S.S.R. 1965;**7**:1768-1776

[62] Yağci Y, Reetz I. Externally stimulated initiator systems for cationic polymerization. Progress in Polymer Science. 1998;**23**:1485-1538

[63] Ueno K, Yamaoka H, Hayashi K, Okamura S. Studies on radiation-induced ionic polymerization—II. Effect of solvent on the polymerization of isobutene at low temperature. The International Journal of Applied Radiation and Isotopes. 1966;**17**(10):595-602

[64] Tsuda Y. Radiation-induced ionic polymerization at low temperature. Journal of Polymer Science. 1961;**54**(159):193-199

[65] Sigwalt P, Moreau M. Carbocationic polymerization: Mechanisms and kinetics of propagation reactions. Progress in Polymer Science. 2006;**31**(1):44-120

[66] Matyjaszewski K. Activated esters in the cationic polymerization of styrenes. In: Makromolekulare Chemie. Macromolecular Symposia; Vol. 13; No. 1; Hüthig & Wepf Verlag; 1988. pp. 433-441

[67] Matyjaszewski K, Pugh C. In: Matyjaszewski K, editor. Cationic Polymerizations: Mechanisms, Synthesis, and Application (Plastics Engineering 35). New York: Marcel Dekker, Inc; 1996. pp. 189-225

[68] Clough RL. High-energy radiation and polymers: A review of commercial processes and emerging applications. Nuclear Instruments and Methods in Physics Research Section B: Beam Interactions with Materials and Atoms. 2001;**185**(1):8-33

[69] Mincher BJ, Wells RP, Reilly HJ. Public Acceptability of the Use of Gamma Rays from Spent Nuclear Fuel as a Hazardous Waste Treatment Process. Idaho Falls, ID, United States: EG and G Idaho, Inc.; 1992

[70] Chmielewski AG, Haji-Saeid M, Ahmed S. Progress in radiation processing of polymers. Nuclear Instruments and Methods in Physics Research Section B: Beam Interactions with Materials and Atoms. 2005;**236**:1, 44-54

[71] Okamura S, Hayashi K, Kitanishi Y. Radiation-induced solid-state polymerization of ring compounds. Journal of Polymer Science Part A: Polymer Chemistry. 1962;**58**(166):925-953

[72] Baysal B, Adler G, Ballantine D, Colombo P. Solid state polymerization of acrylamide initiated by gamma radiation. Journal of Polymer Science Part A: Polymer Chemistry. 1960;**44**(143):117-127

[73] Cheng J, Deming TJ. Synthesis of polypeptides by ring-opening polymerization of α-amino acid N-carboxyanhydrides. In: Peptide-Based Materials. Berlin Heidelberg: Springer; 2011. pp. 1-26

[74] Erdmenger T, Guerrero-Sanchez C, Vitz J, Hoogenboom R, Schubert US. Recent developments in the utilization of green solvents in polymer chemistry. Chemical Society Reviews. 2010;**39**(8):3317-3333

Chapter 7

Interaction with Matter of Ionizing Radiation and Radiation Damages (Radicals)

Betül Çalişkan and Ali Cengiz Çalişkan

Additional information is available at the end of the chapter

http://dx.doi.org/10.5772/intechopen.74691

Abstract

Interaction of matter with "ionizing radiation," that is, high-energy electromagnetic radiation (X- or gamma rays) or α- or β-particles, can promote chemical change which commonly involves free radicals. Free radicals formed by high-energy radiation in solids can then be identified by analysis of their EPR spectra.

Keywords: ionizing radiation, radical, radiation chemistry, radiation damage center, electron paramagnetic resonance (EPR)

1. Introduction

The interaction of ionized radiation with matter brings some changes in matter. These changes are called the radical or the radiation damage centers. Among the methods of radical formation in matter structure, ionizing radiation plays an important role. In addition to ionizing radiation, the chemical reactions and the thermal effects can also lead to radical formation. There are three different applications for radical formation:

a. irradiation (photolysis and radiolysis),

b. thermolysis (thermal homolysis),

c. redox processes (oxidation-reduction reactions) [1, 2].

In irradiative processes, the energy required to form radicals is generated either by **electromagnetic radiation (ultraviolet (UV) or visible light, X-rays, etc.)** or by **particle radiation** (high-energy electrons; α, β, γ particles; neutrons; protons; etc.). Processes involving light absorption are called photochemistry [1]. Deterioration of the structure of the material by light effect is called photolysis. Deterioration of the structure of the material exposed to ionizing radiation is called radiolysis.

The electron paramagnetic resonance (EPR) method is used for the investigation of the radiation damage centers, which occur in the structure of the material. The breakage in the structure of the material allows the material to become paramagnetic. The paramagnetic centers are determined by EPR spectroscopy.

Much work has been done on the interaction of organic and inorganic materials with ionizing radiation. Most of these studies are on EPR [3, 4]. Generally, EPR works with gamma rays, which have the highest energy radiation [5–7].

The aim of the chapter is to recognize the types of radiation, to group the charged or uncharged particles in the ionized radiation group, and to focus on the EPR method, which will examine their effect on matter. High-energy energetic gamma and X-rays cause permanent damage to the material. They constitute defects in chemical structure. These disorders are often referred to as paramagnetic centers or radiation damage centers. Radiation damage centers are seen as free radicals in the form of breaking bonds in the structure and as anion or cation radicals in the form of electron exchange. The EPR method is used to investigate such impairments or paramagnetic centers occurring in structures of organic or inorganic materials. In this chapter, the effects of ionizing radiation on the substance will be analyzed using the EPR method.

2. Interaction with matter of ionizing radiation and radiation damages (radicals)

2.1. Ionizing radiation

Ionizing or high-energy radiation devices can be examined in three groups [8–10]:

a. charged particle accelerators (X-ray generators, medium-energy electron accelerators, and Van der Graaff accelerator),

b. radioisotope sources (^{60}Co beam source, ^{137}Cs beam source, and spent fuel elements),

c. nuclear reactor (mixed radiation of gamma quantum and neutrons). Since the ^{60}Co beam source is easily controlled [11, 12], it is more preferred.

Reactive radicals can be formed by the effect of ionizing radiation [13]. Radiation breaks bonds between atoms in molecules. The deterioration of the bonds occurs in two ways. If the bonds are broken **homolytically (paramagnetic), free radicals** will form. If the bonds are broken **heterolytically (diamagnetic), ionic fragments** are formed [14].

Certain types of low-energy radiation, such as ultraviolet light, may also cause ionization. Such radiation is not from the ionizing radiation group. The lowest energy limit for ionizing radiation is around 10 keV.

Charged particles are a **directly ionizing radiation** group. These particles include **energetic electrons (negatrons), positrons, protons, alpha particles, charged mesons, muons**, and **heavy ions (ionizing atoms)**. This type of ionizing radiation interacts primarily with matter via the Coulomb force. Particles push or attract electrons from atoms and molecules due to their charge.

Uncharged particles form an **indirectly ionizing radiation** group. Best examples for indirectly ionizing radiation type are **photons above 10 keV (X-rays** and **gamma rays**) and **all neutrons**.

Interaction of X-ray and gamma ray photons with matter causes ionization. This process is mainly in three different ways:

a. **Low-energy photons** give all their energy to an electron, which causes the electrons to break apart from atoms or molecules. This type of interaction is called **photoelectric interaction**. The photon then disappears from the medium.

b. **Intermediate-energy photons** interact with electrons as a result of the **Compton event**. Photons and electrons are scattered in different directions.

c. **Photons with more than 1.02 MeV of energy** are affected by pair production. (At around 1.02 MeV, the Compton effect continues, but the pair production is more effective.) As a result, the photon disappears and an electron-positron pair forms (due to momentum and energy conservation, this event occurs near a nucleus). The total kinetic energy of the electron-positron couple is equal to the photon energy, which is lower than the sum of and the rest-mass energies of the electron and the positron (1.02 MeV). These electrons and positrons act as **directly ionizing radiation**. As the positron continues its path, it loses its kinetic energy. As a result, they encounter an electron and destroy each other (the pair disappears). As a result of the pair production, two gamma photons come to the foreground. The two photons, which usually have an energy of 0.511 MeV, spread out to form an angle of 180° with each other.

A photon can enter into interaction with any of these events. However, more than 1022 MeV of energetic photons are required for double formation. The determination of the type of interaction depends on the energy of the photon and the material the photon interacts with.

Interaction of the material with the neutron causes inelastic collisions. These are listed as neutron capture (or activation) and fission. Both interactions are nuclear interactions. A nucleus that collides inelastically with the neutron is excited at a higher energy level. This energy is thrown from the nucleus in three ways:

a. as a gamma ray,

b. as a beta particle,

c. as both a gamma ray and a beta particle.

In the neutron capture, the interacting nucleus absorbs the neutron. After all, the energy from the nucleus goes away in four ways:

a. as a gamma ray,

b. as an X-ray,

c. as a beta particle,

d. by publishing both the gamma ray/X-ray and the beta particle.

Secondary particles cause ionization. If a heavy nucleus absorbs the neutron, then **the fission event** occurs. This may result in two lighter nuclei with radioactive character.

The radiation weighting factor (w_R) is a number that represents the value of **the relative biological effectiveness (RBE)** of the radiation. **The w_R values** are related to **linear energy transfer (LET)** and are given in **Table 1**.

The relative biological activity (RBE) of a radiation type is defined as the inverse ratio of radiation absorption doses that produce biological effects at the same amount. The stochastic biological effect is a biological effect that is caused by ionizing radiation. The higher the dose, the greater the effect. However, this increase does not follow a certain proportion with the absorbed dose. It is a totally independent biological effect. An important example of the stochastic biological effect is cancer.

The tissue weighting factor (w_T) represents the total loss in tissue or organs due to the stochastic effect caused by irradiation of the whole body. Irritation damages tissues or organs. A uniform equivalent dose applied over the entire body should be equal to the number of doses administered separately for all tissues and organs of the body [15].

2.2. Interaction with matter of ionizing radiation and radiation damages (radicals)

The separation of chemical bonds with radiation is one of the most common and direct methods that can be used to produce molecular fragments and free radicals. Ultraviolet, X- or, gamma

Type and energy range	w_R
Photons, all energies	1
Electrons and muons, all energies[2]	1
Neutrons, energy <10 keV	5
10–100 keV	10
>100 keV–2 MeV	20
>2–20 MeV	10
>20 MeV	5
Protons, other than recoil protons, energy >2 MeV	5
Alpha particles, fission fragments, heavy nuclei	20

[1]All values relate to the radiation incident on the body or, for internal sources, emitted from the source.
[2]Excluding Auger electrons emitted from nuclei bound to DNA.

Table 1. Radiation weighting factors w_R[1].

rays can break the bonds between atoms. The minimum energy required depends on the strength of the bond. The bond strength is usually about 10 eV. This value corresponds approximately to the wavelengths in the ultraviolet region. Therefore, UV region is the lowest-energy region required for free radical formation.

X- or gamma ray radiation occurs at very high values. When many solids in amorphous or crystalline form are exposed to X- or gamma rays, a large number of bonds can be broken. Thus, the free radical spectrum can be seen. Such high-energy radiation often produces **high free radical concentrations**. The electron spin (paramagnetic) resonance spectra of the radicals resulting from the high-energy irradiation have high signal-to-noise ratios as a function. However, it is difficult to say that there is a precise mechanism as a free radical production method, because there are many different types of damage.

In all irradiation experiments, it is usually necessary to use the capture mechanism to collect a large concentration of radicals. Examination of the dynamic concentration of radicals in the solution by X-ray or UV irradiation should normally be possible. The concentration at any time is determined by a balance between the rate of formation and the recombination rate. X- or gamma ray irradiation provides sufficient kinetic energy to move away from the site of molecular fragment or broken bonds at a certain speed. Thus, molecular fragments and broken bonds are trapped in the solid to prevent recombination at a sufficient distance. As a result, radicals can occur only in a viscous or a rigid solid phase and can be trapped. In ultraviolet irradiation, fragments or radicals obtained have a high rate of recombination since they do not have excess kinetic energy after bond breaking. Therefore, trapping is less likely to occur.

The results obtained by ultraviolet irradiation method will be discussed first. X- or gamma ray irradiation method will be examined in terms of "organic" and "inorganic" compounds. It should be assumed that the radiation of interest completely releases the chemical bond and leaves two molecules of an electron containing an unpaired electron. Apart from bond breakage, the bond excitation state deals with the "triplet state" which is guided in the same direction, instead of breaking into a higher-energy level by reversing one of the bond electrons.

2.2.1. Ultraviolet irradiation

Ingram et al. [16] obtained the first electron paramagnetic resonance spectra of active radicals formed by ultraviolet irradiation and trapped at low temperatures. In the experiments, the compounds to be photolyzed are dissolved in various hydrocarbons and frozen to form solid glasses at liquid nitrogen temperatures. The viscosity of the resulting glass is an important parameter. After formation, there must be a structure with little rigidity that will not prevent the radicals from moving away from where they are. On the other hand, it must have a solid structure at the level that will prevent the deterioration of radical structure. Glass can be used at room temperature for such a trapping operation. Such experiments, which lead to the formation of stable radicals in plastic films, have been performed by Bijl and Rose-Innes [17].

The glass-trapping technique at low temperature was used in the studies by Lewis and Lipkin [18], which study reactive molecules, and Norman and Porter [19], who detect free radicals trapped by UV absorption spectroscopy. However, it has been shown that the glass required for UV is also suitable for electron resonance studies, since the small defects that will scatter the UV wavelengths badly have no effect on the microwave radiation.

2.2.1.1. *Secondary radicals*

In such experiments, it has been observed that secondary radicals occur frequently when radicals are produced by photolysis by reacting with solvent molecules [20]. Hydroxyl radicals are derived from **hydrogen peroxide**. These radicals are very reactive. When a small amount of hydrogen peroxide is added to the hydrocarbon glass, secondary radicals are usually obtained by proton removal from the hydrocarbons. **OH radicals** were obtained by removal of an α-hydrogen from the isopropanol molecule via UV irradiation.

Hydrogen atom reactions and high irradiation are predictors of **secondary reaction products** for solid hydrocarbons [16]. EPR is also an ideal metric for the analysis of these complex radicals. For example, the H atoms in the frozen aqueous solutions of $\mathbf{H_2SO_4}$, $\mathbf{H_3PO_4}$, or $\mathbf{HClO_4}$ present at liquid nitrogen temperatures (77 K) are quite stable. This provides excellent environments for future work [21].

2.2.2. *X and gamma ray irradiation*

2.2.2.1. *Organic compounds*

Radiation of an organic material can lead to breakage of bonds between atoms. The minimum energy required for such a situation can be determined by the bond strength [21, 22]. Schneider et al. [23] observed the first electron spin (paramagnetic) resonance spectrum of radicals obtained from the organic polymethylmethacrylate compound by X-irradiation. They produced a very large electron spin (paramagnetic) resonance spectrum giving a complex hyperfine structure splitting. With the detailed analysis of the EPR spectrum, it is understood that the actually occurring state is the result of two separate interaction mechanisms in the $R - CH_2 - \dot{C}(CH_3)(COOCH_3)$ radical [24, 25].

Gordy et al. [26, 27] have begun to investigate the radicals of biological substances exposed to X-radiation. As a clear result of the radiation damage, they found that the hyperfine splits originate from simple radical forms. The organic materials that gave these spectra were the various irradiated amino acids. Their EPR spectra were obtained after irradiation with 50 kV X-rays at room temperature. The triplet was obtained from glycine, $CH_2(NH_2)COOH$. The hyperfine interaction ion is due to a $\dot{C}H_2$ radical having two protons. X-rays or other high-energy radiation can break down the molecules and create a variety of radicals. There are many ways to create radicals. One of them is to remove an electron, which is the primary action of irradiation, and to form an ionized molecule with small stable groups such as NH_3 CO_2 and $(CH_2)^+$ fragments of glycine [26].

It has been observed that the EPR spectrum of the irradiated alanine, $CH_3.CH(NH_2).COOH$, has five symmetrical lines. This suggests that the unpaired electron interacts with four equivalent protons and that the spin density is in a $(H_2C - CH_2)^+$ group due to hyperconjugation [26].

It has been observed that the EPR spectrum obtained from the irradiated valine, $CH_3.CH(CH_3).CH(NH_2).COOH$, is composed of five and seven overlapping lines. The seven-line splitting was assumed to originate from a $(H_3C-CH_3)^+$ radical group and was found to be larger than that obtained from UV-irradiated isopropyl alcohol [20]. This may be due to the partial electron withdrawal of the OH group in the structure giving the seven-line splitting.

However, as in the case of UV-irradiated spectra and in stable aromatic ions, the theories of hyperfine interaction and hyperconjugation reveal these spectra in detail. Line widths are the same as those obtained from frozen UV-irradiated glasses. Thus, it can be argued that motional narrowing in these systems is due to dipolar interaction in the nearest proton.

Gordy et al. [26, 27] studied various amino, carboxylic, and hydroxy acids that have been X-irradiated. In some studies, large symmetric doublets are presumably due to the dipole-dipole interaction between the unpaired electron localized on an oxygen or a sulfur atom and a single proton near it. They also investigated the effect of temperature on the carboxylic acid spectra. The irradiated acetic acid is a good example of this. The hydrogen bond keeps the molecules very hard at 77 K. Thus, it has been assumed that the doublet occurs from the dipole-dipole interaction between an electron and a hydroxyl proton localized on hydroxyl oxygen [27].

Gordy et al. [26] obtained spectra obtained from X-radiation of various biological materials such as proteins and bone tissue. Comparing these spectra with those obtained from simpler acids, one can get an idea of the deterioration mechanisms in natural products. Uebersfeld et al. [28–30] studied sugar, cellulose, and similar materials exposed to gamma rays in detail. In their later work, they obtained an anisotropic EPR spectrum from a single crystal of glycine [31]. A similar situation has been observed in the study of single crystals of the X-irradiated alanine by Van Roggen et al. [32]. Whiffen et al. [25, 33] have also done much research on the analysis of EPR spectra from gamma-irradiated polymers with interesting "**oxygen effects.**".

DNA is an organic compound. The EPR spectrum of the radical formed in gamma-irradiated DNA is shown in **Figure 1** [34]. As another organic compound, potassium hydroquinone monosulfonate (PHM) may be mentioned. In potassium hydroquinone monosulfonate single crystal, EPR spectrum of the radical structure resulting from gamma irradiation is shown in **Figure 2** [35].

2.2.2.2. Inorganic compounds

Irradiation of **inorganic materials**, usually in a single crystal form, produces centers of damage trapped within the crystal lattice. This can form free atoms, molecules, or radicals in the crystal

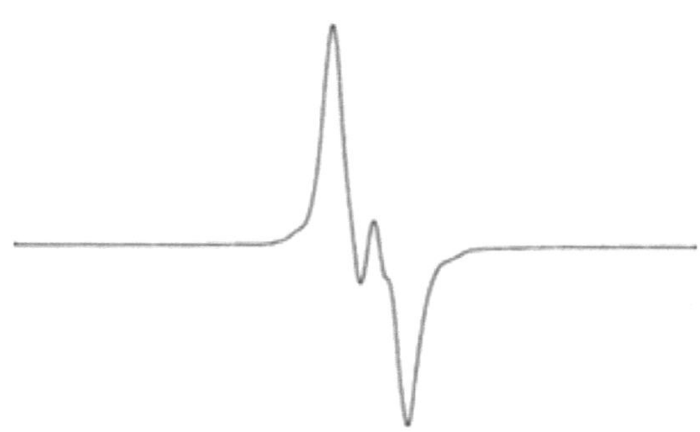

Figure 1. The EPR spectrum of the gamma-irradiated DNA at 77 K.

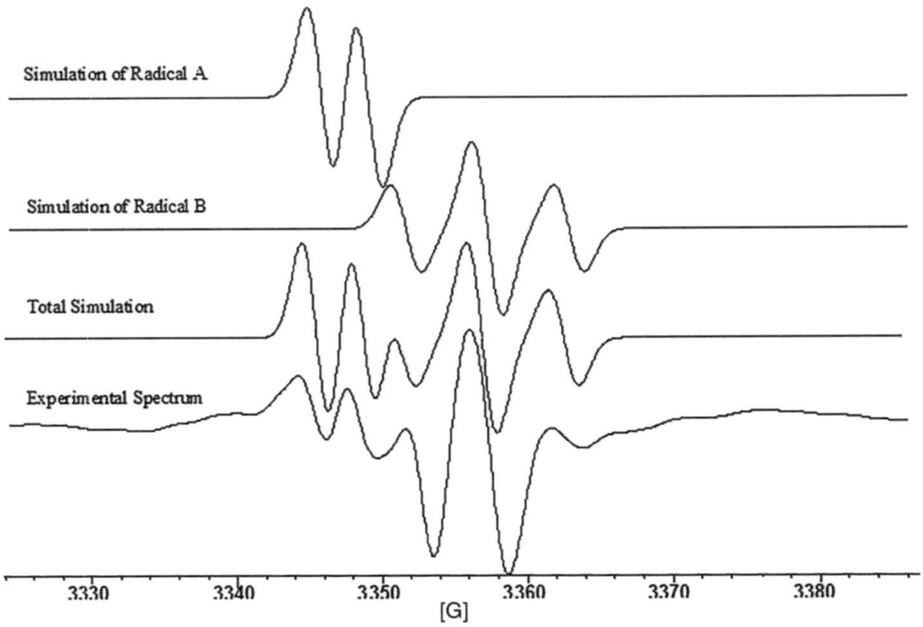

Figure 2. EPR spectrum of gamma-irradiated PHM single crystal at 125 K.

as well as breakage of chemical bonds [21]. The paramagnetic entities produced by the irradiation in these alkali halide crystals can hardly be termed "**free radicals**" in the normal sense of the word. They are better classified as "**damage centers**" or "**defects**" since they are associated with the structure of the crystal itself, rather than a broken bond in a particular molecule.

The EPR study of irradiated inorganic compounds was made before organic samples. Hutchison [36] is the first person to work on this subject. He studied the EPR spectra of lithium fluoride crystals irradiated with neutrons for several hours. This work was followed by X-irradiation studies by researchers such as Schneider and Britain [37], Tinkham and Kip [38], and Kip et al. [39]. In these alkali halide crystals, it is more accurate to denote radiation-generated paramagnetic objects as "**damage centers**" or "**defects**" rather than "**free radicals.**"

Calcite is an inorganic material. The EPR spectrum of the radical structure of gamma-irradiated calcite is shown in **Figure 3** [40].

The damage centers generated by high-energy irradiation in inorganic single crystals can be examined in three categories:

a. *F*-centers,

b. *V*-centers,

c. *U*-centers (Interstitial atoms).

2.2.2.2.1. F-centers

Negative ion deficiency (a halogen vacancy, an anion vacancy) in the lattice and **trapping of an unpaired electron (trapped electron) in the Coulomb field** of the vacancy create **an**

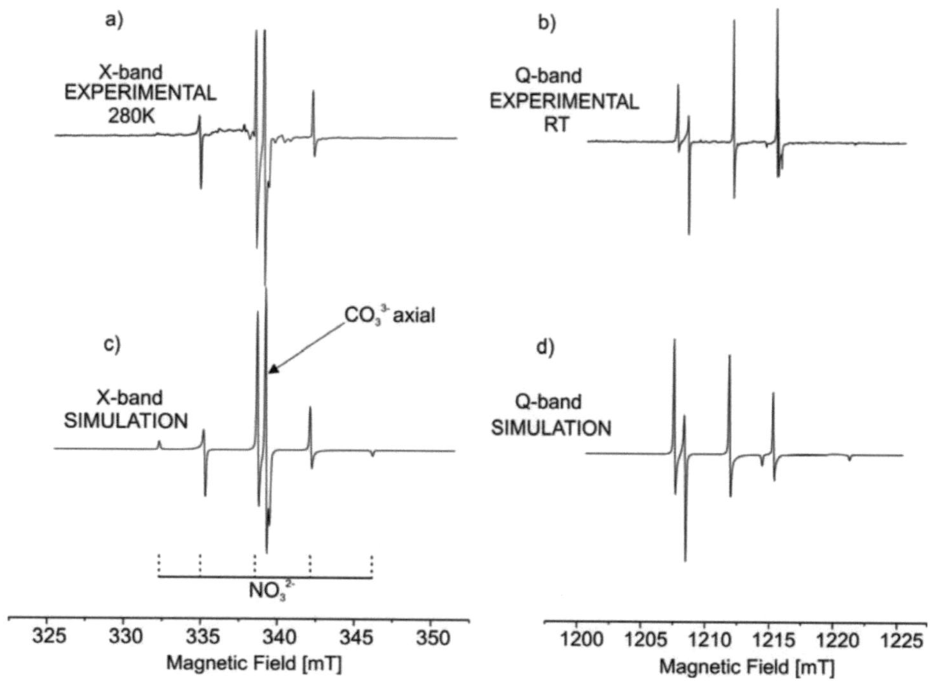

Figure 3. EPR spectra of gamma-irradiated calcite at 77 K recorded at room temperature: (a) X-band spectrum; (b) Q-band spectrum; (c) the simulation of the X-band spectrum; and (d) the simulation of the Q -band spectrum.

***F*-center**. The unpaired electron also moves on neighboring ions. Previous results of F-centers have also been confirmed by Feher's double resonance experiments [41]. Thanks to this technique, all the hyperfine splitting of the *F*-center in **KCl** can be solved and the coupling constants can be measured in detail. Resolved hyperfine structure has been examined by Lord [42], who studied the *F*-centers formed in **LiF** and **NaF**. Many authors have studied the molecular orbital interaction theory of the unpaired electron of the *F*-center [39, 43–50].

EPR spectra of the X-irradiated $LiBaF_3$ crystal along the [100], [111], and [110] directions are given in **Figure 4**. EPR spectra of F-centers formed according to the main orientations of the $LiBaF_3$ crystal were clearly observed [51].

2.2.2.2.2. *V-centers*

The opposite of **an *F*-center** is **a *V*-center**. **A positive ion deficiency (a cation vacancy) in the lattice** and **a deletion of electrons (trapped hole center) in one of the neighboring ions** form **a *V*-center**. Thus, **this hole**, which is scattered over orbital neighbor ions, is **trapped in vacancy**. When two such ***V*-centers** are generated side by side, **a V_2-center** is obtained. In this case, the spins of the two electron holes are doubled and the paramagnetism ceases to exist. **A V_3-center** can be defined as **a V_2-center** with only one missing electron or hole between both vacancies. There is an unpaired spin like **a V_1-center**. Kanzig made initial measurements on *V*-centers [52]. V-centers were obtained by X-irradiation of **KCl** at 90 K. A g-value was found between 2.0023 and 2.024 and was observed to be anisotropic. Anisotropic situation has occurred due to the fact that the hyperfine splitting originating from the two chlorine

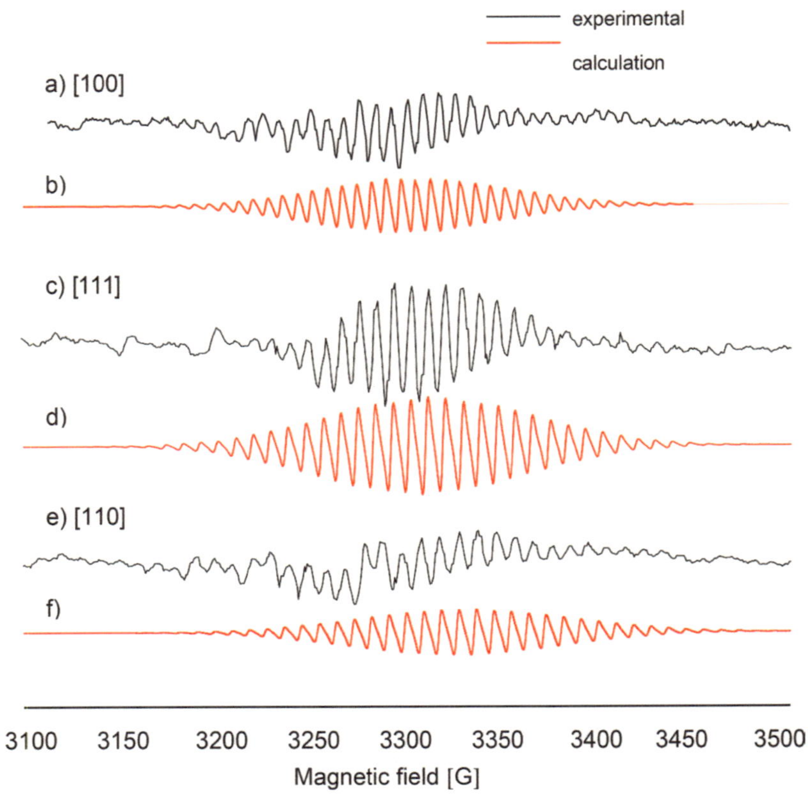

Figure 4. EPR spectra of the F-centers in the LiBaF3 crystal.

nuclei (Cl_2^- ion) shows an angle-dependent change. For this reason, the vacancies are defined as V_3-centers rather than V_1-centers, because V_3-centers must have a central cubic symmetry [53]. The results are consistent with the measurements made on X- and gamma-irradiated and **thallium-activated KCl** [54]. The observed g-value can also be calculated very well by molecular orbital theory [55].

EPR spectra of defects occurring in ZnO with ultraviolet effect are given in **Figure 5**. These defects are the oxygen-vacancy and Zn-vacancy centers [56].

2.2.2.2.3. *Interstitial atoms and U-centers*

Generally, **interstitial atoms (*U*-centers)** are obtained when an atom of high-energy radiation, such as neutrons, is sent to an intermediate position from a normal lattice site. For this reason, adding a single atom to the normal ionic bond of the lattice leaves a vacant site. The centers of damage expressed as ***F*-center** and ***V*-center** occur in the case of **pure single crystals**. However, if there is **hydride** or **deuteride in the lattice**, it is possible to create ***U*-centers** defined as **negative ion vacancies containing a hydrogen or a deuterium negative ion**. Thus, the unpaired electron is localized on hydrogen and brings a large doublet hyperfine splitting. The best example of a defect generated by interstitial atoms is **neutron-irradiated diamond**. Griffiths et al. [57] studied the electron spin (paramagnetic) resonance

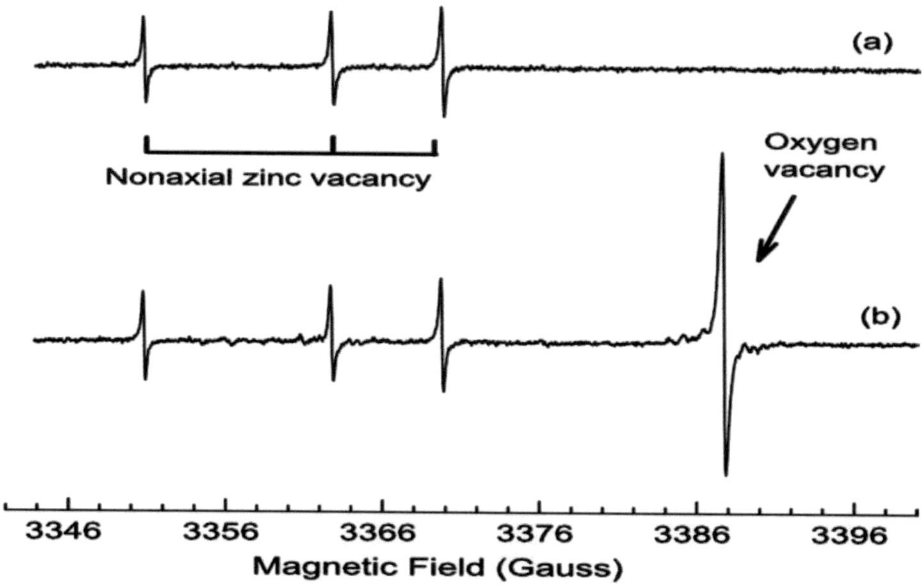

Figure 5. EPR spectra of ZnO **(a)** in the dark and **(b)** after illumination with 325-nm light.

spectrum of this defect and obtained a strong central line with very weak satellites. While the central line [58] forms single interstitial carbon atoms, the satellites defined by magnetic centers also bring about double interstitial aggregates forming C_2 molecules. Wertz et al. [59] found that the trapping centers were produced at the same time as the interstitial atoms during their work on **neutron-irradiated magnesium oxide**. In the vacancies created by neutron irradiation, electrons are trapped and a well-resolved hyperfine structure splitting is obtained from neighboring Mg nuclei [41]. Delbecq et al. [54] observed ***U*-centers** in **irradiated mixed crystals of KCl-KH** and **KCl-KD**. In this study, it is seen that the unpaired electron interacts with the negative ion vacancy and has a doublet of 500 G for **hydrogen** and a triplet of 78 G for **deuterium**.

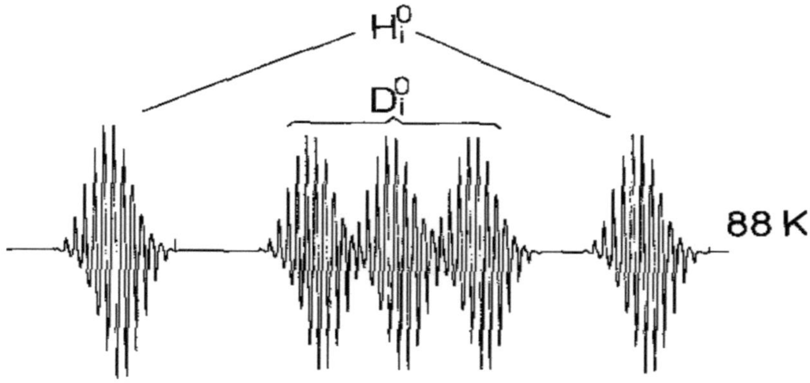

Figure 6. The EPR spectrum of $H\,i\,0$ (U_2 -center) and $D\,i\,0$-centers into (LiH) +.

At low temperatures below 70 K, interstitial atoms are produced by ultraviolet or X-ray irradiation. These atoms form in hydrogenated or hydroxyl-doped crystals. In the KCl, the H_i^0 -center (U_2-center) and D_i^0-center are observed due to the ultraviolet effect [60]. In KCl, the $(LiH)^+$-center is transformed to produce H_i^0-center (U_2-center) and D_i^0-center. The EPR spectrum of the resulting H_i^0-center (U_2-center) and D_i^0-center is shown in **Figure 6**.

3. Conclusion

In addition to chemical reaction and thermolysis methods, a material is subjected to a series of changes by irradiation. Photolysis or radiolysis processes are used as irradiation methods. However, radiolysis is more effective. Radiolysis is a process performed using ionizing radiation. Ionized radiation is separated into charged and uncharged particles. In this group, X-rays and gamma rays are used more effectively than neutrons. Interactions of such high-energy noncharged particle radiation or indirectly ionizing radiation with matter are effective in breaking chemical bonds or ionizing the structure. Breaking of the structure and its ionic conversion are expressed as free radicals and ionic radicals, respectively.

When organic and inorganic substances are exposed to ionizing radiation, the bond structures deteriorate. In organic and inorganic materials, UV, X-, or gamma rays can give rise to ionic radicals or free radicals. The radiation damage centers in inorganic materials can be seen as *F*-center, *V*-center, and *U*-center. In inorganic pure single crystals, *F*- and *V*-centers are named as radiation damage centers, negative ion deficiency, and positive ion deficiency, respectively. As the radical lifetimes can be very short, the detection of the radical becomes difficult. To overcome such a difficulty, EPR analysis is often performed at low temperature. The life of the radical in the sample, which is kept at low temperature, is extended.

In EPR studies, irradiation of organic or inorganic material is the first step. The type of radiation used must be highly energized. Using gamma rays leads to more accurate results. The second step is to perform EPR analysis at low temperature (77 K). Thus, the single electron formed by breaking the bonds is provided to be trapped at the point where it exists. At low temperature, the removal of the radical from the environment and the damping of the radical are prevented. This provides the appropriate time for the analysis of the radical.

Author details

Betül Çalişkan[1]* and Ali Cengiz Çalişkan[2]

*Address all correspondence to: bcaliska@gmail.com

1 Department of Physics, Faculty of Arts and Science, Pamukkale University, Denizli, Turkey

2 Department of Chemistry, Faculty of Science, Gazi University, Ankara, Turkey

References

[1] Pryor WA. Free Radicals. New York: McGraw-Hill Book Company; 1966

[2] Symons M. Chemical and Biochemical Aspects of Electron-Spin Resonance Spectroscopy. New York: Van Nostrand Reinhold Company; 1978. pp. 94-96

[3] Caliskan B. EPR study of gamma irradiated cholestanone single crystal. Acta Physica Polonica A. 2014;**125**(1):135-138. DOI: 10.12693/APhysPolA.125.135

[4] Caliskan B, Caliskan AC, Yerli R. Electron paramagnetic resonance study of radiation damage in isonipecotic acid single crystal. Journal of Molecular Structure. 2014;**1075**:12-16. DOI:10.1016/j.mol.struc.2014.06.030

[5] Caliskan B, Caliskan AC. EPR study of radiation damage in gamma irradiated 3-nitro-acetophenone single crystal. Radiation Effects and Defects in Solids. 2017;**172**(5-6):398-410. DOI: 10.1080/10420150.2017.1320800

[6] Caliskan B, Caliskan AC, Er E. Electron paramagnetic resonance study of radiation-induced paramagnetic centers in succinic anhydride single crystal. Journal of Molecular Structure. 2017;**1144**:421-431. DOI: 10.1016/j.molstruc.2017.05.039

[7] Caliskan B, Caliskan AC. EPR study of free radical in gamma-irradiated bis(cyclo-pentadienyl)zirconium dichloride single crystal. Radiation Effects and Defects in Solids. 2017;**172**(5-6):507-516. DOI: 10.1080/10420150.2017.1346652

[8] Charlesby A, editor. Radiation Sources. New York: McMillan; 1964

[9] Dole M, editor. Radiation Chemistry of Macromolecules. Vol. 1. New York: Academic Press; 1972. p. 265

[10] Johnsen RH. Ionizing radiation, instrumentations and methods. In: Lamola AA, Ware WR, editors. Creation and Detection of the Excited State. New York: Marcel Dekker; 1971. p. 429

[11] Ranby B, Rabek JF. ESR Spectroscopy in Polymer Research. Springer-Verlag, Berlin; 1977. pp. 7-10

[12] Roylance DK. An investigation of polymer fracture [doctorate thesis]. 1968. p. 173

[13] Perkins MJ. Radical Chemistry. New York: Ellis Horwood Limited; 1994. pp. 29-30

[14] Bersohn M, Baird JC. An Introduction to Electron Paramagnetic Resonance. New York: W.A. Benjamin, Inc., 1966. pp. 190-193

[15] Cherry RN Jr. Introduction, Chapter 48-Radiation: Ionizing. In: Cherry RN Jr, Chapter editor. Part VI - General Hazards. Online Edition: Encyclopaedia of Occupational health and Safety (http://www.iloencyclopaedia.org/)

[16] Ingram DJE, Hodgson WG, Parker CA, Rees WT. Detection of labile photochemical free radicals by paramagnetic resonance. Nature. 1955;**176**:1227-1228. DOI: 10.1038/1761227a0

[17] Bijl D, Rose-Innes AC. Preparation of solid solutions of free radicals at room temperature. Nature. 1955;**175**(4445):82-83. DOI: 10.1038/175082b0

[18] Lewis GN, Lipkin D. Reversible photochemical processes in rigid media: The dissociation of organic molecules into radicals and ions. Journal of the American Chemical Society. 1942;**64**(12):2801-2808. DOI: 10.1021/ja01264a025

[19] Norman I, Porter G. Trapped atoms and radicals in rigid solvents. Proceedings of the Royal Society of London. Series A. 1955;**230**(1182):399-414. DOI: 10.1098/rspa.1955.0142

[20] Gibson JF, Ingram DJE, Symons MCR, Townsend MG. Electron resonance studies of different radical species formed in rigid solutions of hydrogen peroxide after UV irradiation. Transactions of the Faraday Society. 1957;**53**:914-920. DOI: 10.1039/TF9575300914

[21] Assenheim HM. Introduction to Electron Spin Resonance. New York: Plenum Press; 1967. pp. 141-143

[22] Ingram DJE. Free Radicals as Studied by Electron Spin Resonance. London: Butterworths Scientific Publications; 1958. pp. 170-171

[23] Stein G, Day MJ, Schneider EE. Effects of X-rays upon plastics: Paramagnetic resonance. Nature. 1951;**168**:645. DOI: 10.1038/168645a0

[24] Ingram DJE, Symons MCR, Townsend MG. Electron resonance studies of occluded polymer radicals. Transactions of the Faraday Society. 1958;**54**:409-415. DOI: 10.1039/TF9585400409

[25] Abraham RJ, Melville HW, Ovenall DW, Whiffen DH. Electron spin resonance spectra of free radicals in irradiated polymethyl methacrylate and related compounds. Transactions of the Faraday Society. 1958;**54**:1133-1139. DOI: 10.1039/TF9585401133

[26] Gordy W, Ard WB, Shields H. Microwave spectroscopy of biological substances. I. Paramagnetic resonance in X-irradiated amino acids and proteins. Proceedings of the National Academy of Sciences of the United States of America. 1955;**41**(11):983-996

[27] Gordy W, Ard WB, Shields H. Microwave spectroscopy of biological substances. II. Paramagnetic resonance in X-irradiated carboxylic and hydroxy acids. Proceedings of the National Academy of Sciences of the United States of America. 1955;**41**(11):996-1004

[28] Uebersfeld JCR. On the multiplicity of the bands of paramagnetic resonance in certain irradiated organic substances. Academy Science (Paris). 1954;**239**:240-241

[29] Uebersfeld J. A paramagnetic resonance spectrometer. Study of free radicals in irradiated substances (γ-rays) and coals. Annals of Physics. Series 13. 1956;**1**:395-461

[30] Schneider EE, Bleaney B, Tinkham M, Ingram DJE, Runciman WA, Wieringen JSv, George P, Griffith JS, Pake GE, Combrisson J, Uebersfeld J, Gordy W, Dainton FS, Livingston R,

Zeldes H, Taylor EH, Extemann CR, Denis P, Béné G, Townes CH. General discussion. Discussions of the Faraday Society. 1955;**19**:173-186. DOI: 10.1039/DF9551900173

[31] Uebersfeld J, Erb E. Anisotropy of the electronic paramagnetic resonance spectra in irradiated glycine. Comptes Rendus. 1956;**242**:478-480

[32] Van Roggen A, Van Roggen L, Gordy W. Electron-spin resonance in x-irradiated single crystals of amino acids. Bulletin of the American Physical Society. 1956;**1**:266-267

[33] Abraham RJ, Ovenall DW, Whiffen DH. Electron resonance spectrometer for the investigation of organic free radicals. Transactions of the Faraday Society. 1958;**54**:1128-1132. DOI: 10.1039/TF9585401128

[34] Becker D, Bryant-Friedrich A, Trzasko C, Sevilla MD. Electron spin resonance study of DNA irradiated with an argon-ion beam: Evidence for formation of sugar phosphate backbone radicals. Radiation Research. 2003;**160**(2):174-185. DOI: 10.1667/RR3037

[35] Caliskan B, Caliskan AC, Er E. Electron paramagnetic resonance study of gamma-irradiated potassium hydroquinone monosulfonate single crystal. Radiation Effects and Defects in Solids. 2016;**171**(5-6):440-450. DOI: 10.1080/10420150.2016.1203924

[36] Hutchison CA. Paramagnetic resonance absorption in crystals colored by irradiation. Physical Review. 1949;**75**(11):1769-1770. DOI: 10.1103/PhysRev.75.1769.2

[37] Schneider EE, England TS. Paramagnetic resonance at large magnetic dilutions. Physica. 1951;**17**(3-4):221-233. DOI: 10.1016/0031-8914(51)90062-6

[38] Tinkham M, Kip AF. Paramagnetic resonance absorption crystals containing color centers. Physical Review. 1951;**83**(3):657-658. DOI: https://doi.org/10.1103/PhysRev.83.657.2

[39] Kip AF, Kittel C, Levy RA, Portis AM. Electronic structure of F centers: Hyperfine interactions in electron spin resonance. Physical Review. 1953;**91**(5):1066-1071. DOI: 10.1103/PhysRev.91.1066

[40] Sadło J, Bugaj A, Strzelczak G, Sterniczuk M, Jaegermann Z, Multifrequency EPR. Study on radiation induced centers in calcium carbonates labeled with ^{13}C. Nukleonika. 2015;**60**(3):429-434. DOI: 10.1515/nuka-2015-0076

[41] Feher G. Electronic structure of F centers in KCl by the electron spin double-resonance technique. Physical Review. 1957;**105**(3):1122-1123. DOI:10.1103/PhysRev.105.1122

[42] Lord NW. Hyperfine structure of F-center spin resonance in lithium fluoride and sodium fluoride. Physical Review. 1957;**105**(2):756-757. DOI:10.1103/PhysRev.105.756

[43] Muto T. Theory of the F-centers of coloured alakali halides. Part II: Electronic structure of F-centers – General theory. Progress of Theoretical Physics. 1949;**4**(3):243-250. DOI: 10.1143/ptp/4.3.243

[44] Inui T, Uemura Y. Theory of color centers in ionic crystals. I – Electronic states of F-centers in alkali halides. Progress of Theoretical Physics. 1950;**5**(2):252-265. DOI: 10.1143/ptp/5.2.252

[45] Inui T, Uemura Y. Theory of color centers in ionic crystals. II – The temperature dependence of F-bands in alkali halides. Progress of Theoretical Physics. 1950;**5**(3):395-404. DOI: 10.1143/ptp/5.3.395

[46] Kahn AH, Kittel C. F-center wave functions and electronic g-values in KCl crystals. Physical Review. 1953;**89**(1):315-315. DOI: 10.1103/PhysRev.89.315

[47] Kittel C. Microwave resonance and the electronic structure of colour centers and impurity states in solids. Report of the Conference on Defect in Crystalline Solids. Physical Society at University of Bristol. 1955; London: Physical Society of London; 1955. p. 33

[48] Dexter DL. F-center wave functions in alkali halides. Physical Review. 1954;**93**(1):244-245. DOI: 10.1103/PhysRev.93.244.2

[49] Krumhansl JA. Theoretical calculations of F-center energy levels. Physical Review. 1954; **93**(1):245-246. DOI: 10.1103/PhysRev.93.245

[50] Seitz F. Color centers in alkali halide crystals. II. Reviews of Modern Physics. 1954;**26**(1): 7-94. DOI: 10.1103/RevModPhys.26.7

[51] Fedotovs A. EPR of radiation defects in fluoride crystals and in oxyfluoride glass ceramics. [PhD thesis]. Riga: University of Latvia Faculty of Physics And Mathematics; 2008. pp. 34-37

[52] Känzig W. Electron spin resonance of V_1-centers. Physical Review. 1955;**99**(6):1890-1891. DOI: 10.1103/PhysRev.99.1890

[53] Cohen MH. Spin resonance in V-centers. Physical Review. 1956;**101**(4):1432-1433. DOI: 10.1103/PhysRev.101.1432

[54] Delbecq CJ, Smaller B, Yuster PH. Paramagnetic resonance investigation of irradiated KCl crystals containing *U*-centers. Physical Review. 1956;**104**(3):599-604. DOI:10.1103/PhysRev.104.599

[55] Inui T, Harasawa S, Obata Y. Note on the electron spin resonance of the V_1-center – Calculation of the electronic g-value. Journal of the Physical Society of Japan. 1956; **11**(5):612-613. DOI: 10.1143/JPSJ.11.612

[56] Kappers LA, Gilliam OR, Evans SM, Halliburton LE, Giles NC. EPR and optical study of oxygen and zinc vacancies in electron-irradiated ZnO. Nuclear Instruments and Methods in Physics Research Section B. 2008;**266**(12-13):2953-2957. DOI: 10.1016/j.nimb.2008.03.146

[57] Griffiths JHE, Owen J, Ward IM. Magnetic resonance in irradiated diamond and quartz. Report of the Conference on Defect in Crystalline Solids. Physical Society at University of Bristol. 1954; London: Physical Society of London; 1954. p. 81-87

[58] O'Brien MCM, Pryce MHL. Paramagnetic resonance in irradiated diamond and quartz: Interpretation. Report of the Conference on Defect in Crystalline Solids. Physical Society at University of Bristol. 1954; London: Physical Society of London; 1954. p. 88

[59] Wertz JE, Auzins P, Weeks RA, Silsbee RH. Electron spin resonance of F centers in magnesium oxide; confirmation of the spin of magnesium-25. Physical Review. 1957; **107**(6):1535-1537. DOI: 10.1103/PhysRev.107.1535

[60] Schwan LO, Geigle W, Paus H. Optical and EPR investigations of a new hydrogen center in potassium chloride. Zeitschrift für Physik B. 1979;**35**(1):43-50. DOI: 10.1007/BF01322080

Recent Developments in Count Rate Processing Associated with Radiation Monitoring Systems

Romain Coulon and Jonathan Dumazert

Additional information is available at the end of the chapter

http://dx.doi.org/10.5772/intechopen.71233

Abstract

This chapter presents some recent data processing developments associated with radiation monitoring systems. Radiation monitors have to continuously provide count rate estimations with accuracy and precision. A filtering technique based on a Centered Significance Test coupled with a Brown's double exponential filter has been developed and used in compensation measurement and moving sources detection schemes.

Keywords: data processing, nuclear counting, radiation monitor, signal processing, filtering, frequentist inference

1. Introduction

During the last decades, ionizing ray detectors have grown in performance, thanks to digital electronics developments (ADC and FPGA), allowing for an advanced processing of nuclear impulse signals. It is also noteworthy that this field has favored the development of real-time processing algorithms dealing with count rate data.

The architecture of a typical nuclear measurement system is presented in **Figure 1**. It can be divided in four parts:

- Voltage supply,
- Detector part,
- Front-end electronics,
- User interface.

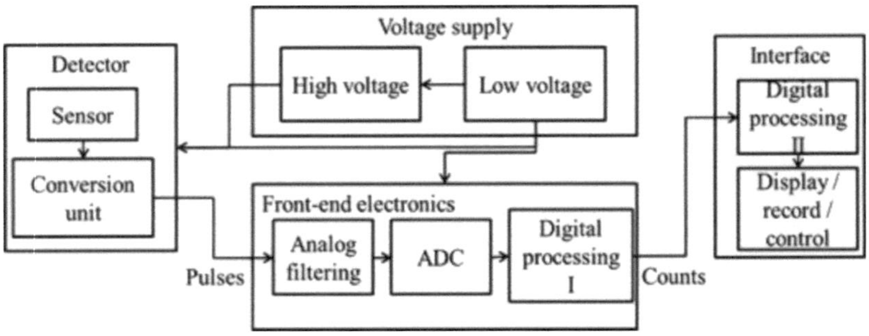

Figure 1. Schematic architecture of a nuclear measurement apparatus.

The detector part contains the physical sensor (noble gas, scintillation material, and semiconductor) in which radiation interacts with matter. A conversion unit (preamplifier or photo-converter) converts the induced charges or photons in amplified voltage pulses. In the case of gas or semiconductor-based sensors, a high voltage is required to polarize the medium, and a low voltage is needed to supply active components of the preamplifier. In the case of scintillators, a high voltage supplies the photomultiplier.

Front-end electronics is composed of an analog filter, an analog-to-digital converter (ADC), and a digital filter. An analog shaping filter can be used to adapt the signal before digital conversion (dynamic range and respect to Shannon rules), and/or to maximize signal-to-noise ratio (SNR). The ADC digitalizes the signal with a given frequency and resolution.[1] This digital signal is processed into a fast electronic component, typically a microcontroller, or a field-programmable gate array (FPGA). The embedded firmware has to comply with the very high-frequency of the ADC output with a processing period in the range of 1–10 ns. The algorithms implemented in the firmware perform the pulse processing, which mainly consists in triggering, first digital filtering for SNR maximization, stabilizing the baseline, estimating the dead time, and counting a number of pulse events N over a period of time $\Delta\tau$. This general description is not exhaustive and a variety of architectures is conceivable, depending on the mix between analog and digital processing. Though modern trends tend to favor digital filtering, analog filtering can still be retained to comply with cost reduction or embedded strategies. For instance, the front-end associated with a scintillator can directly digitalize the output voltage of a photomultiplier using a 500 MS/s ADC and process the pulses using a FPGA (notably when pulse shape discrimination is needed). On the other hand, the initial signal can be filtered using analog components (trapezoidal filtering), before digitalization with a 10 MS/s ADC and count processing with a microcontroller.

An interface is built on a computer connected with the front-end electronic card. The software reads, at each given time interval Δt, a new count value N_i according to a defined communication protocol. This second processing can be divided in two parts: filtering of the count rate signal and displaying. The period Δt has to be chosen in compliance with continuous measurement requirement, typically close to retinal persistence of 0.1 s. We can highlight here, the

[1]Current ADCs are available with tradeoffs between resolution and sampling frequency such as: 16 bit / 100MS/s and 8 bit / 1 GS/s. CAEN Electronic instrumentation, 724 Digitizer Family, CEAN data sheet, 2015.

quantitative difference between the digital pulse processing coded in VHDL or Verilog for the FPGA (very fast), and the count rate processing which can be coded in C/C++ into a microcontroller or/and the PC interface (7–8 orders of magnitude slower). In compact systems, the count rate processing is usually incorporated into the firmware while, in larger systems, the count rate processing is remotely implemented in the PC.

This chapter will not address pulse processing techniques, for which details can be found in [1–3], but presents some recently developed techniques to process count rate signal using frequentist inference. Bayesian inference can also be implemented to process count rate as for instance for gamma spectrum unfolding or photon-limited imaging filtering [4, 5]. These are very efficient to accurately processed nuclear counting data, but become unsuited of online applications. After describing the theoretical model of the counting process, a smoothing technique will be presented as a fundamental building block, ensuring an online and adaptive filtering of the signal. The issue of composite measurements will then be addressed with a method allowing improving metrological reliability for particle discrimination (compensation technique). Finally, the use of detectors in a network to address moving source detection will be developed.

2. Nuclear counting model

Nuclear disintegration can occur following different processes depending on the A/Z ratio of the concerned isotope. Major disintegration processes read: β^-, β^+, ε, α, and spontaneous fission decay, presented in the following nuclear equations, where X is the mother nucleus and Y the daughter nucleus

$$ {}_{Z}^{A}X \xrightarrow{T_{1/2}} {}_{Z+1}^{A}Y^* + {}_{-1}^{0}\beta^{-1} - + {}_{0}^{0}\overline{\upsilon}_e \tag{1} $$

$$ {}_{Z}^{A}X \xrightarrow{T_{1/2}} {}_{Z-1}^{A}Y^* + {}_{1}^{0}\beta + + {}_{0}^{0}\upsilon_e \tag{2} $$

$$ {}_{Z}^{A}X + {}_{-1}^{0}e^- \xrightarrow{T_{1/2}} {}_{Z-1}^{A}Y^* + {}_{0}^{0}\upsilon_e \tag{2a} $$

$$ {}_{Z}^{A}X \xrightarrow{T_{1/2}} {}_{Z-2}^{A-4}Y^* + {}_{2}^{4}\alpha \tag{3} $$

$$ {}_{Z}^{A}X \xrightarrow{T_{1/2}} {}_{Z1}^{A1}Y_1{}^* + {}_{Z1}^{A1}Y_2{}^* + k_0^1 n \tag{4} $$

Subsequently, the daughter nucleus is, most of the time, released in an exited state and usually reaches its fundamental level by gamma-ray emission:

$$ {}_{Z}^{A}Y^* \rightarrow {}_{Z}^{A}Y + {}_{0}^{0}\gamma \tag{5} $$

According to the detector type, β^-, β^+, α, n, or γ particles are detected and counted. The required time τ_d for an unstable nucleus to decay is undetermined, and takes its value in an exponential distribution whatever the time lap between its creation and the observation is (memoryless phenomenon). The probability distribution $p(\tau_d = t')$ of the decay instant, where t' and λ are, respectively, the observation instant and the decay constant of the nucleus, is given by:

$$p(\tau_d = t') = \lambda \exp(-\lambda t') \tag{6}$$

The observation of an unstable nucleus over a time t forms a Bernoulli trial in which two results can be observed: the nucleus has decayed or the nucleus has not decayed. A probability of p and $1 - p$ can be, respectively, associated with each branch of the trial for every instant as illustrated in **Figure 2**.

The probability $p_{X \rightarrow Y}(t)$ to observe a disintegration of the mother nucleus $X \rightarrow Y$ before time t ($t' = 0$ being the start of the observation) is obtained as:

$$p_{X \to Y}(t) = \int_0^t p(\tau_d = t')dt' = 1 - e^{-\lambda t} \tag{7}$$

In a radioactive source containing a population of N_X unstable nuclei, the decay of an individual nucleus does not impact the decay of the others. The Bernoulli trial is therefore repeated N_X times (**Figure 3**) during the observation time t, and the number of observed decays n is described by a Binomial law such as:

$$p(N_{X \to Y}(t) = n) = \frac{N_X!}{(N_X - n)!n!} p_{X \to Y}(t)^n \left(1 - p_{X \to Y}(t)\right)^{N_X - n} \tag{8}$$

In practice, N_X is very large and $p_{X \rightarrow Y}(t)$ is usually very small ($1/\lambda \gg t$). In these conditions, the Binomial law converges toward a Poisson law $\mathcal{P}$ such as:

$$p(N_{X \to Y}(t) = n) = \mathcal{P}(N_X \lambda t) = \frac{(N_X \lambda t)^n}{n!} e^{-N_X \lambda t} \tag{9}$$

Expectation and variance of the number of decays are equal to $N_X \lambda t$. The number of counts N measured before observation time t is obtained by weighting Eq. (9) with the detection efficiency ε and the probability η of the detected particle to be emitted during decay. The expected count rate $\rho = \varepsilon \eta N_X \lambda$ thus becomes the parameter of the distribution of measured count values before t:

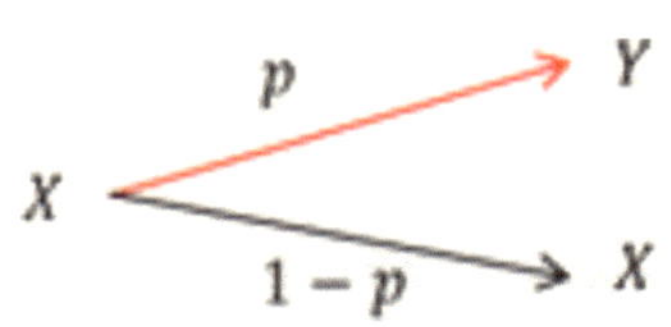

Figure 2. Illustration of the Bernoulli trail applied to an individual nucleus disintegration.

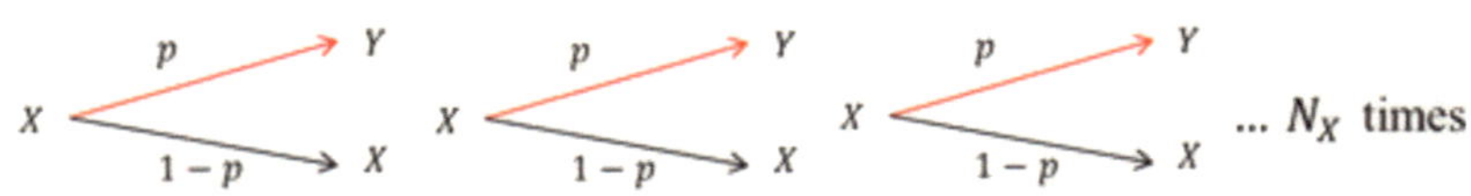

Figure 3. Illustration of Bernoulli trials applied to a population of unstable nuclei.

$$p(N(t) = n) = \mathcal{P}(\rho t) = \frac{(\rho t)^n}{n!} e^{-\rho t} \tag{10}$$

At each time t_i, sampled such as $t_0 = 0$ & $\forall i \geq 1, t_i = t_{i-1} + \Delta t$, the raw estimation of the count rate $\rho(t_i) = \rho_i$ is provided by measuring N_i, which is a time-dependent random variable taking its values in a Poisson distribution such as:

$$N_i \sim \mathcal{P}(\rho_i \Delta t) \tag{11}$$

A challenge in radiation monitoring is to provide count rate estimation ρ_i at each time t_i maximizing both precision $\sigma(\rho_i) \to 0$ and accuracy $\sigma(t_i) \to 0$. Algorithmic techniques to meet this expectation are discussed in the next section.

3. Count rate smoothing

The aim of smoothing algorithms is to improve the estimation of ρ_i, originally defined as: $\widehat{\rho}_i = \frac{N_i}{\Delta t}$. This improvement can be achieved by using past values $N_{i-1}, N_{i-2}, \ldots$ recorded in a memory according to the assumption that $p(\rho_i | N_i, N_{i-1}, N_{i-2}, \ldots)$ is more precise than $p(\rho_i | N_i)$. If we consider, in a first approach, a constant count rate ρ, the estimator which maximizes the likelihood of a homogenous Poisson process is the average [6]:

$$\widehat{\rho}_i = \frac{1}{(m+1)\Delta t} \sum_{j=i-m}^{i} N_j \tag{12}$$

where $m+1$ is the temporal depth of the filter and ϑ is a kernel function in which each $\vartheta_{j, 1 \leq j \leq i}$ equal to one. According to the property of equality between variance and expectation, the associated variance $\sigma^2(\widehat{\rho}_i)$ can be estimated as:

$$\sigma^2(\widehat{\rho}_i) = \frac{1}{(m+1)\Delta t} \sum_{j=i-m}^{i} N_j \tag{13}$$

The relative stochastic uncertainty $\sigma(\widehat{\rho}_i)/\widehat{\rho}_i$ is inversely proportional to the square root of the historical depth $m+1$.

In practice, counting processes are not homogenous (ρ is not constant). In this case, it is important to provide an estimate of the time $\widehat{t}_i$ effectively corresponding to the current count rate estimate $\widehat{\rho}_i$. Because the sampling times t_i over the temporal depth $m+1$ are identically weighted, the estimate $\widehat{\rho}_i$ from Eq. (12) is associated to a time estimate $\widehat{t}_i = t_i - \frac{m+1}{2}\Delta t$ with a temporal precision $\sigma(\widehat{t}_i) = \frac{(m+1)\Delta t}{2\sqrt{3}}$. We therefore see that $\sigma(\widehat{\rho}_i)$ can only be minimized to the detriment of $\sigma(\widehat{t}_i)$, leading to a degradation of accuracy when ρ is varying. One way to address this issue is to actualize the temporal depth m_i after every count rate estimation. The optimal value for m_i is a function of the temporal behavior of the rate ρ at the time t_i: $\left.\frac{d\rho}{dt}\right|_{t=t_i} = 0$ or $\left.\frac{d\rho}{dt}\right|_{t=t_i} > 0$.

First approaches consist in the implementation of preset count filter providing a fixed variance $\sigma^2(\widehat{\rho}_i)$, or finite impulse response (FIR) filters in which a kernel function ϑ is used to assign more weight to recent than older count values such as Eq. (12) become:

$$\widehat{\rho}_i = \frac{\sum_{j=i-m}^{i} \vartheta_{i-j} N_j}{(m+1)\Delta t \sum_{j=1}^{m} \vartheta_j} \tag{14}$$

Among FIR filters, the exponential moving average (EMA) remains widespread [7, 8], but do not fully deals with the tread-off issue between accuracy and precision.

The algorithm translation of the actualization of m_i is the building of infinite impulse response (IIR) dedicated to nuclear counting [9]. Such nonlinear filtering requires a hypothesis test to detect the changes in count rate ρ. The null hypothesis $\mathbf{H_0}$ and the detection hypothesis $\mathbf{H_1}$ are formalized as follows:

$$\mathbf{H_0}\text{: } \forall j \in [i - m_i; i], \rho_j = \theta_0 \tag{15}$$

$$\mathbf{H_1}\text{: } \exists j \in [i - m_i; i], \rho_j = \theta_1 \tag{16}$$

In a first approach [10], a sequential probability ratio test (SPRT) has been assessed under the assumption that θ_1 is a known value. Later, generalized tests in which θ_1 is an unknown parameter have been introduced, notably the generalized likelihood ratio test (GLR) [11] and the centered significance test (CST) [12]. In these change detection algorithms, several estimations of the current count rate are calculated using different temporal depths k such as:

$$\widehat{\rho}_i^k = \frac{1}{(k+1)\Delta t} \sum_{j=i-k}^{i} N_j \tag{17}$$

In the rest of the discussion, we will conventionally use notation $\widehat{\rho}_i^k$ to designate both the underlying random variable and its actual values.

In the CST test, the vector $\widehat{\rho}_i^{k, 1 \leq k \leq m_i}$ is scanned to find a potential change in the true rate ρ. For every temporal depth $k \in [\![1; m_i]\!]$, the difference between count rate estimations $\Delta\widehat{\rho}_i^k = \widehat{\rho}_i^{m_i} - \widehat{\rho}_i^k$ is the quantity which will be tested for significance.

The method is based on a comparison between actual and expected distributions of $\Delta\widehat{\rho}_i^k$ under H_0 and H_1, respectively [13]. The distribution $\mathfrak{D}$ of $\Delta\widehat{\rho}_i^k$ is the difference between two weighted Poisson distributions

http://dx.doi.org/10.5772/intechopen.71233

$$\Delta\widehat{\rho}_i^k \sim \mathfrak{D} = \left\{\frac{1}{(m_i+1)\Delta t}\mathcal{P}\left(\widehat{\rho}_i^{m_i}\,(m_i+1)\Delta t\right) - \frac{1}{(k+1)\Delta t}\mathcal{P}\left(\widehat{\rho}_i^k(k+1)\Delta t\right)\right\} \tag{18}$$

The expectation $\mathbb{E}\left(\Delta\widehat{\rho}_i^k\right) = \mathbb{E}\left(\widehat{\rho}_i^{m_i}\right) - \mathbb{E}\left(\widehat{\rho}_i^k\right) = \Delta\theta$ between times $(i-m_i)\Delta t$ and $i\Delta t$. Moreover, we will make use of assumption $\Delta\widehat{\theta} = \widehat{\rho}_i^{m_i} - \widehat{\rho}_i^k$, as common in nuclear counting experiments will finite statistics [1]. The variances associated with both uncorrelated random processes are summed to obtain a cumulative standard deviation for $\Delta\widehat{\rho}_i^k$. According to the equality between expectation and variance, we obtain:

$$\sigma\left(\Delta\widehat{\rho}_i^k\right) = \sqrt{\frac{\sigma^2\left(\widehat{\rho}_i^{m_i}(m_i+1)\Delta t\right)}{((m_i+1)\Delta t)^2} + \frac{\sigma^2\left(\widehat{\rho}_i^k(k+1)\Delta t\right)}{((k+1)\Delta t)^2}} \approx \sqrt{\frac{\widehat{\rho}_i^{m_i}}{(m_i+1)\Delta t} + \frac{\widehat{\rho}_i^k}{(k+1)\Delta t}} \tag{19}$$

We will note $\mathfrak{D}\left(\mathbb{E}\left(\Delta\widehat{\rho}_i^k\right), \sigma\left(\Delta\widehat{\rho}_i^k\right)\right)$, the distribution of the difference random variable with its first and second order moments.

Under $\mathbf{H_0}$, $\mathbb{E}\left(\Delta\widehat{\rho}_i^k\right) = 0$ (*cf.* left curve in **Figure 4**). A decision threshold (*DT*) is determined in compliance with a given risk of false detection $\alpha_i^k = p(\mathbf{H_1}|\mathbf{H_0})$. *DT* is defined in the following formula, where $Q_{1-\alpha_i^k}$ is the quantile of the error function (err) with a confidence level $1-\alpha_i^k$:

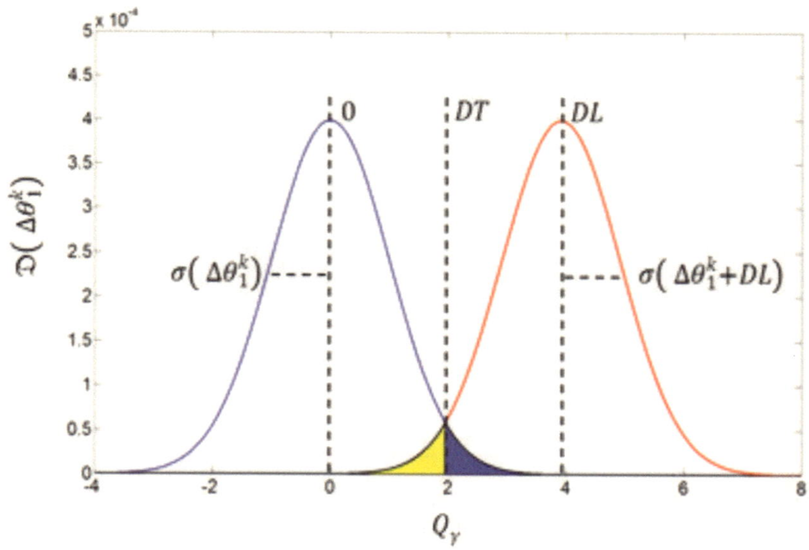

Figure 4. Illustration of distributions $\mathfrak{D}_{H_0}$ (left curve) and $\mathfrak{D}_{H_1}$ (right curve) and construction rules of the hypothesis test.

$$\alpha_i^k = \int_{DT}^{\infty} \mathfrak{D}_{H_0}\left(0, \sigma\left(\Delta\widehat{\rho}_i^k\right)\right) d\Delta\widehat{\rho}_i^k \approx err\left(Q_{1-\alpha_i^k}\right) \tag{20}$$

In practice, for embedded implementations, it is impossible to sample and interpolate distributions $Q_{1-\alpha_i^k} = f\left(\alpha_i^k\right)$ for every value of i and k. Moreover, when $\mathbb{E}\left(\widehat{\rho}_i^{m_i}\right)$ and $\mathbb{E}\left(\widehat{\rho}_i^k\right)$ are large enough, distribution $\mathfrak{D}_{H_0}$ may be approximated as $\mathcal{N}\left(0, \sigma\left(\Delta\widehat{\rho}_i^k\right)\right)$, with $\mathcal{N}$ the Normal law. Under this assumption, for every value of i and k, $\alpha_i^k = \alpha$ and $Q_{1-\alpha_i^k} = Q_{1-\alpha}$, where $Q_{1-\alpha}$ is a quantile of $\mathcal{N}$ and err becomes:

$$err(Q_{1-\alpha}) = 2 - 2\Phi(Q_{1-\alpha}) \tag{21}$$

where Φ is the cumulative distribution function of the centered Normal law.

As illustrated in **Figure 4**, DT can be calculated thanks to the weighting of the standard deviation by $Q_{1-\alpha}$ such as:

$$DT_i^k = Q_{1-\alpha}\, \sigma\left(\Delta\widehat{\rho}_i^k\right) \tag{22}$$

If $\Delta\widehat{\rho}_i^k \leq DT_i^k$, hypothesis $\mathbf{H_0}$ is accepted with a confidence level equal to $1-\alpha$.

Under $\mathbf{H_1}$, $\mathfrak{D}_{H_1}\left(\mathbb{E}\left(\Delta\widehat{\rho}_i^k\right), \sigma\left(\Delta\widehat{\rho}_i^k\right)\right) \approx \mathcal{N}\left(DL_i^k, \sigma\left(\Delta\widehat{\rho}_i^k + DL_i^k\right)\right)$ (cf. right curve in **Figure 4**), where DL is defined as the detection limit. DL is determined in compliance with a given risk of non-detection $\beta = p(\mathbf{H_0} \mid \mathbf{H_1})$. DL is obtained in the following formula:

$$\beta = \int_{-\infty}^{DT} \mathcal{N}\left(DL_i^k, \sigma\left(\Delta\widehat{\rho}_i^k + DL_i^k\right)\right) d\theta = err\left(Q_{1-\beta}\right) \tag{23}$$

As illustrated in **Figure 4**, DL can be calculated thanks to the weighting of the associated standard deviation by the quantile $Q_{1-\beta}$ of the error function erf such as:

$$DL_i^k = DT_i^k + Q_{1-\beta}\, \sigma\left(\Delta\widehat{\rho}_i^k + DL_i^k\right) \tag{24}$$

An equivalent confidence level $1-\alpha = 1-\beta = \gamma$ is considered, and Eq. (23) is solved recursively such as:

$$\forall y \in [1; \infty],$$

$$DL_{i,y}^k = Q_\gamma\left[\sigma\left(\Delta\widehat{\rho}_i^k\right) + \sigma\left(\Delta\widehat{\rho}_i^k + DL_{i,y-1}^k\right)\right] \tag{25}$$

With

$$DL_{i,0}^{k} = 2Q_{\gamma}\sigma\left(\Delta\widehat{\rho}_{i}^{k}\right) \tag{26}$$

When $y \to \infty$,

$$DL_{i}^{k} = Q_{\gamma}^{2} + 2Q_{\gamma}\,\sigma\left(\Delta\widehat{\rho}_{i}^{k}\right) \tag{27}$$

If $\Delta\widehat{\rho}_{i}^{k} \geq DL_{i}^{k}$, hypothesis $\mathbf{H_1}$ is accepted with a confidence level $1-\beta=\gamma$.

The number L_i of significant changes recorded into memory $\Delta\widehat{\rho}_{i}^{k}$ is used to calculate the next value of temporal depth m_{i+1}:

$$L_i = \dim\left\{\underset{k,\,1\leq k\leq m_i}{\arg}\left[\left|\Delta\widehat{\rho}_{i}^{k}\right| > DL_{i}^{k}\right]\right\} \tag{28}$$

If $L_i=0$, true rate ρ is considered to remain constant and historical depth may be extended $m_{i+1}=m_i+1$, to the benefit of a reduction of $\sigma(\widehat{\rho})$ (better precision). On the other hand, if $L_i>0$, true rate ρ is considered to change and the historical depth needs to be reduced $m_{i+1}=m_i-L_i$, to the benefit of $\sigma\left(\widehat{t}\right)$ (better accuracy).

At every elementary time step Δt, the retained count rate estimate $\widehat{\rho}_{i}^{*}$ is therefore calculated over an adaptable temporal depth m_i, Eq. (17) becoming:

$$\widehat{\rho}_{i}^{*} = \frac{1}{(m_i+1)\Delta t}\sum_{j=i-m_i}^{i} N_j \tag{29}$$

With

$$\sigma^2\left(\widehat{\rho}_{i}^{*}\right) = \frac{1}{((m_i+1)\Delta t)^2}\sum_{j=i-m_i}^{i} N_j \tag{30}$$

This nonlinear approach performs advantageously in comparison with conventional linear filters [12, 14], allowing to maintain sufficient precision while rate changes in the signal occur.

Remaining high-frequency fluctuations can now be reduced using a second, recursive smoother, for instance a Brown's double exponential filter [14]. A first exponential smoothing $\widehat{\rho}_{i}^{1}$ is performed on $\widehat{\rho}_{i}^{*}$ with a smoothing parameter δ_i such as:

$$\widehat{\rho}_{i}^{1} = \delta_i\widehat{\rho}_{i}^{*} + (1-\delta_i)\widehat{\rho}_{i-1}^{*} \tag{31}$$

With

$$\sigma^2\left(\widehat{\rho}_{i}^{1}\right) = \left[\delta_i\sigma\left(\widehat{\rho}_{i}^{*}\right)\right]^2 + \left[(1-\delta_i)\sigma\left(\widehat{\rho}_{i-1}^{*}\right)\right]^2 \tag{32}$$

A last exponential smoothing $\widehat{\rho}_i^2$ is eventually performed on $\widehat{\rho}_i^1$ under the form:

$$\widehat{\rho}_i^2 = \delta_i \widehat{\rho}_i^1 + (1 - \delta_i)\widehat{\rho}_{i-1}^1 \tag{33}$$

With

$$\sigma^2\left(\widehat{\rho}_i^2\right) = \left[\delta_i \sigma\left(\widehat{\rho}_i^1\right)\right]^2 + \left[(1 - \delta_i)\sigma\left(\widehat{\rho}_{i-1}^1\right)\right]^2 \tag{34}$$

The parameter δ_i changes as a function of the parameter m_i and its strength is set with the parameter W:

$$\delta_i = 1 - \exp\left(-\frac{1}{W(m_i - 1)}\right) \tag{35}$$

Finally, the Brown's estimation $\widehat{\rho}_i^{**}$ is calculated such as:

$$\widehat{\rho}_i^{**} = 2\widehat{\rho}_i^1 - \widehat{\rho}_i^2 \tag{36}$$

With

$$\sigma^2\left(\widehat{\rho}_i^{**}\right) = 4\sigma^2\left(\widehat{\rho}_i^1\right) + \sigma^2\left(\widehat{\rho}_i^2\right) \tag{37}$$

Figures 5 and **6** illustrate the advantage of the hereby described nonlinear filters over conventional, moving average filters with a 20% rate variation, respectively, in a low count rate configuration (5 counts per sample) and in a higher count rate configuration (500 counts per sample). The nonlinear filter has been set with parameters $Q_\gamma = 1.645$ ($\gamma = 90\%$) and $W = 0.2$ and

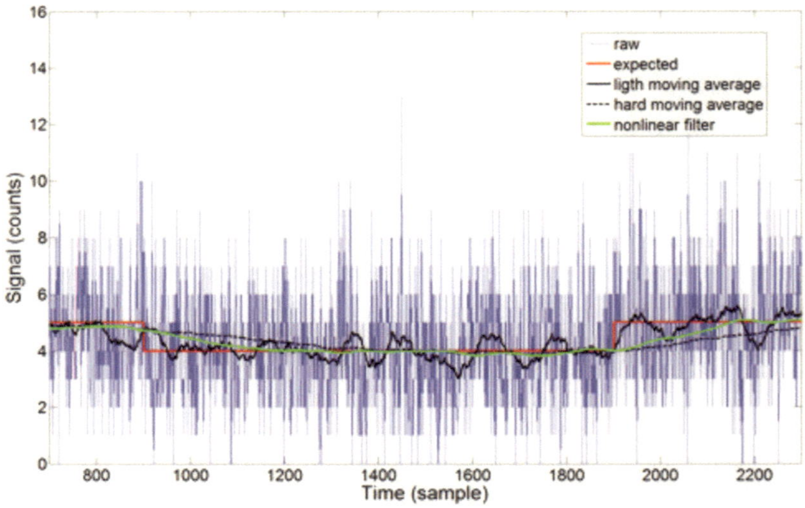

Figure 5. Behavior of smoothing filters over a 20% rate variation at 5 counts per sample.

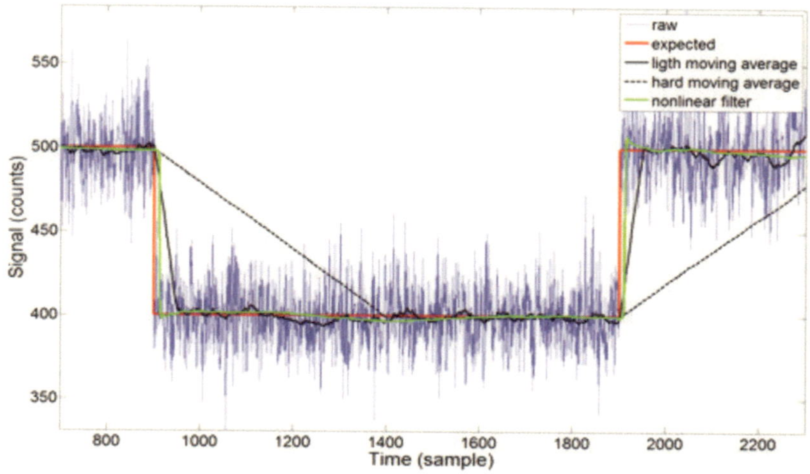

Figure 6. Behavior of smoothing filters over a 20% rate variation at 500 counts per sample.

compared to moving average filters set with $m = 50$ samples (soft) and $m = 500$ samples (hard). **Figure 5** shows that nonlinear filtering offers a better compromise between precision and accuracy, though the detection of small changes within large statistical fluctuation remains unreachable $\left(\Delta\widehat{\rho}_i^k \leq DL_i^k\right)$. At higher count rates (**Figure 6**), nonlinear filtering permits the detection of the rate change $\left(\Delta\widehat{\rho}_i^k > DL_i^k\right)$ and ensures a significant gain, operating both faster and more precisely than both moving averages.

Such nonlinear smoothing algorithms, easily embedded into programmable components, have for instance been implemented into a Geiger-Müller dosimeter fixed on a wireless robot used for radiological threat detection [15]. This algorithmic building block is plays a key role in the nuclear counting methods studied in the next sections, namely compensation measurements and sensor network processing.

4. Compensation measurement

In many cases, radiation monitoring requires the counting of a signal from a first radiation source within an interference signal induced by a second particle emitter, namely α/β vs. γ; n vs. γ, γ vs. γ … The most efficient techniques consist in the recognition of the particle origin associated with each individual pulse event by coincidence/anti-coincidence, pulse height discrimination (PHD) or pulse shape discrimination (PSD) [16]. However, event-by-event discrimination techniques may be found unreliable in particular mixed field configurations. Compensation methods are an alternative solution when addressing such limitations [17, 18]: their principle lies within measuring count rates ρ_A from a first detector A, sensitive to all particles, and comparing the result with count rates ρ_B from a second detector B, only significantly sensitive to background contributions (typically gamma rays). The estimation ρ_C of the count rate associated with particles of interest is obtained by subtraction of ρ_A with ρ_B such as:

$$\rho_C = \rho_A - \omega\rho_B \tag{38}$$

where ω is a correction factor taking in account the fact that detector B is not strictly equivalent to detector A in terms of response as a function of the energy and spatial localization of incident background particles. Three challenges are to be faced in compensation measurement:

- increase in fluctuation level;
- apparition of negative count rates without physical sense;
- loss of reliability (impact of energy and anisotropy of the background signal).

Values of $\rho_{A,i}\Delta t$ and $\rho_{B,i}\Delta t$ at the time t_i are described by Poisson processes, as already stated in the third section of this chapter. Therefore, if $\omega = 1$, values of $\rho_{C,i}\Delta t$ are described by a Skellam process $\mathcal{S}k\left(\rho_{A,i}\Delta t, \rho_{B,i}\Delta t\right)$ such as:

$$\rho_{C,i}\Delta t \sim \mathcal{S}k\left(\rho_{A,i}\Delta t, \rho_{B,i}\Delta t\right) \tag{39}$$

The expectation and the variance of the random variable $\widehat{\rho}_{C,i}\Delta t$ are, respectively, $\mathbb{E}\left(\widehat{\rho}_{C,i}\Delta t\right) = \widehat{\rho}_{A,i}\Delta t - \widehat{\rho}_{B,i}\Delta t = \rho_{A,i}\Delta t - \rho_{B,i}\Delta t$ and $\sigma^2\left(\widehat{\rho}_{C,i}\Delta t\right) = \widehat{\rho}_{A,i}\Delta t + \widehat{\rho}_{B,i}\Delta t = \rho_{A,i}\Delta t + \rho_{B,i}\Delta t$ under the same assumption as in Section 4. The variance definition highlights an increase of fluctuation level in comparison with single-channel measurement. It is therefore required to reduce this variance using a suitable smoothing filter, such as the CST a nonlinear filter described in the previous section (cf. Eqs. (36) and (37)):

$$\left[\widehat{\rho}^{**}_{A,i}; \sigma\left(\widehat{\rho}^{**}_{A,i}\right)\right] = CST\left(\widehat{\rho}_{A,i}\right) \tag{40}$$

$$\left[\widehat{\rho}^{**}_{B,i}; \sigma\left(\widehat{\rho}^{**}_{B,i}\right)\right] = CST\left(\widehat{\rho}_{B,i}\right) \tag{41}$$

Reduced variances $\sigma^2\left(\widehat{\rho}^{**}_{A,i}\right)$ and $\sigma^2\left(\widehat{\rho}^{**}_{B,i}\right)$ calculated according to Eq. (36) are used to determine $\sigma^2\left(\widehat{\rho}^{**}_{C,i}\right)$ as:

$$\sigma^2\left(\widehat{\rho}^{**}_{C,i}\right) = \sigma^2\left(\widehat{\tilde{n}}^{**}_{A,i}\right) + \sigma^2\left(\widehat{\rho}^{**}_{B,i}\right) \tag{42}$$

If the compensation factor ω remains constant but different from 1, Eq. (42) becomes:

$$\sigma^2\left(\widehat{\rho}^{**}_{C,i}\right) = \sigma^2\left(\widehat{\rho}^{**}_{A,i}\right) + \omega^2\sigma^2\left(\widehat{\rho}^{**}_{A,i}\right) \tag{43}$$

In practice, the factor ω is not constant, due to the impact of energy and spatial distributions of incident particles. We then introduce notations $\overline{\omega}$ and $\sigma^2(\omega)$ for the expectation and variance of the variable ω. A resulting variance is therefore calculated by taking into account both statistical error and bias such as:

http://dx.doi.org/10.5772/intechopen.71233

$$\sigma^2\left(\widehat{\rho}_{C,i}^{**}\right) = \sigma^2\left(\widehat{\rho}_{A,i}^{**}\right) + \overline{\omega}^2\sigma^2\left(\widehat{\rho}_{B,i}^{**}\right) + \left(\widehat{\rho}_{B,i}^{**}\right)^2\sigma^2(\omega) \tag{44}$$

The estimation of $\overline{\omega}$ and $\sigma^2(\omega)$ is complicated by the experimental dependence of these parameters. An approach is proposed in [19, 20], in which a database is built from measures acquired in representative areas and in absence of the signal particle of interest: $\rho_C = 0$. In these conditions, compensation factors $\omega_q = \frac{\widehat{\rho}_{A,q}^{**}}{\widehat{\rho}_{B,q}^{**}}$ are obtained for each measurement point ($1 \le q \le n_q$), allowing for an empirical mean $\overline{\omega} = \frac{1}{n_q}\sum_{q=1}^{n_q}\omega_q$ and variance $\sigma^2(\omega) = \frac{1}{n_q}\sum_{q=1}^{n_q}\left(\omega_q - \overline{\omega}\right)^2$ to be estimated.

Based on the generalized variance expressed in Eq. (43), a hypothesis test is built to select positive and significant values of $\widehat{\rho}_{C,i}$. Algorithm 1 presents the detection test in which the presence of particles of interest is detected in compliance with a confidence level γ governing the test. Most of the time, ω_q values can be considered to follow a Normal law, which allows us to apply an envelope coverage factor Q_γ associated with a confident level γ as:

Algorithm 1:

If $\widehat{\rho}_{C,i}^{**} > Q_\gamma\sqrt{2\sigma^2\left(\widehat{\rho}_{C,i}^{**}\right)}$,

Then $\widehat{\rho}_{C,i}^{***} = \widehat{\rho}_{C,i}^{**}$ (detection hypothesis $\mathbf{H}_1$ is accepted)

Else $\widehat{\rho}_{C,i}^{***} = 0$ (detection hypothesis $\mathbf{H}_1$ is rejected)

Figure 7 synthetizes the principle, inputs, and outputs of the compensation technique.

The method improves the reliability of compensation measurement with the use of a recorded database. Moreover, accuracy and decision threshold $Q_\gamma\sqrt{2\sigma^2\left(\widehat{\rho}_{C,i}^{**}\right)}$ associated with the particles of interest are optimized using an adaptive filter, smoothing individual channels while suppressing all negative or non-significant values. The approach described in the present section has been successfully implemented in varied applications, such as α/β contamination meters or gadolinium-based neutron detectors [20, 21].

As a perspective, it has been demonstrated that the multiplication of channels, such as illustrated in **Figure 8**, allows the system to learn a prior distribution for the signal over a set of pixels as a function of incident energy and spatial origin of background particles. Dispatching $\widehat{\rho}_{A,q}^{***}$ and $\widehat{\rho}_{B,q}^{***}$ data along ($X > 1$) dimensions induces a reduction of detection threshold

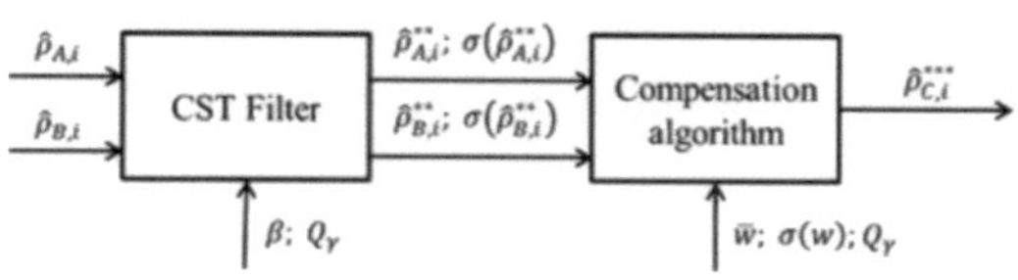

Figure 7. Principle of the compensation technique.

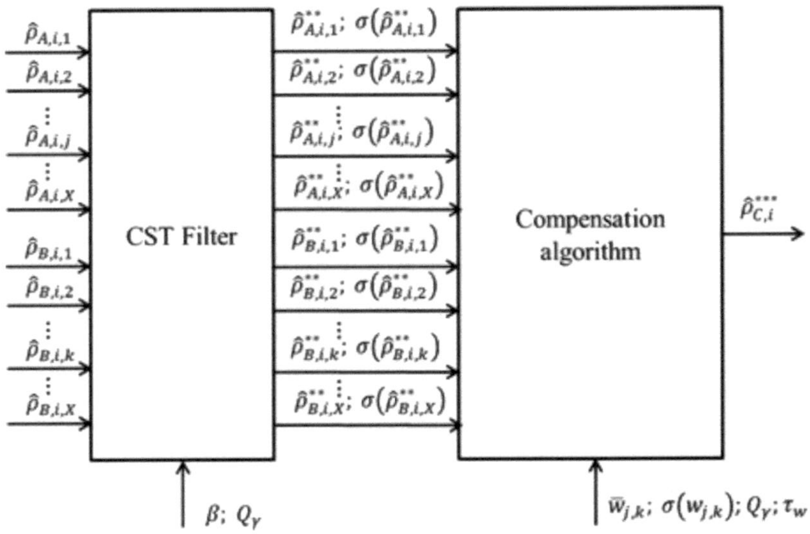

Figure 8. Principle of the compensation technique for pixelated detectors.

$Q_\gamma\sqrt{2\sigma^2\left(\widehat{\rho}^{**}_{C,i}\right)}$ and thus an improvement of measurement reliability. Compensation factors $\overline{\omega}_{j,k}$ and variances $\sigma^2(\omega_{j,k})$ are determined for every $1 \le j \le X$ A-type detector, and every $1 \le k \le X$ B-type detector. Resulting count rates $\rho_{C,i}$ are estimated as:

$$\widehat{\rho}^{**}_{C,i} = \sum_{j=1}^{X}\left[\widehat{\rho}^{**}_{A,i,j} - \sum_{k=1}^{X}\overline{\omega}_{j,k}\widehat{\rho}^{**}_{B,i,k}\right] \tag{45}$$

$$\sigma^2\left(\widehat{\rho}^{**}_{C,i}\right) = \sum_{j=1}^{X}\left\{\sigma^2\left(\widehat{\rho}^{**}_{A,i,j}\right) - \sum_{k=1}^{X}\left[\overline{\omega}_{j,k}\sigma\left(\widehat{\rho}^{**}_{B,i,k}\right)\right]^2 + \left[\widehat{\rho}^{**}_{B,i,k}\sigma(\omega_{j,k})\right]^2\right\} \tag{46}$$

5. Moving source detection

Radiation portal monitors (RPM) are implemented to detect radioactive sources, carried by a vehicle in motion, through the monitoring of a count rate measured by large-volume detectors. Two main issues arise in RPM development: correcting the shadow shielding effect observed when the vehicle is dense enough to impact the baseline of the signal, and improving the detection capability (increasing true detection minus false alarm detection probability).

RPM detection strategy is based on a hypothesis test where the estimated signal $\widehat{\rho}_i$ at the time t_i is continuously compared to a threshold h, itself determined in comparison with the signal distribution under $\mathbf{H_0}$. Let θ_0 be the expected background count rate without any vehicle in the environment surrounding the RPM. A decision threshold (DT) is set, following the same philosophy as presented previously (Eqs. (20)–(26)), as a function of variance $\sigma^2(\theta_0)$ and a confidence level associated with a false detection risk α

$$DT = Q_{1-\alpha}\,\sigma\left(\widehat{\theta}_0 - \theta_0\right) = Q_{1-\alpha}\sqrt{2\sigma^2(\theta_0)} \tag{47}$$

During the passage of a dense vehicle, θ_0 will decrease due to gamma-ray attenuation as the vehicle acts as a radiation shield. Such baseline alteration is noted $\omega\theta_0$, where $\omega \in [0;1]$ is the attenuation factor. An added count rate from a source with intrinsic rate θ_1, put onboard the vehicle, will thus lead to a total signal $\theta_T = \omega\theta_0 + \theta_1$. If $\omega = 1$ and $\theta_1 > DT$, the source is detected with a non-detection risk:

$$\beta_{\omega=1} = err\left[\frac{\theta_1 - DT}{\sigma(\theta_1 + \theta_0)}\right] = err\left[\frac{\theta_1 - Q_{1-\alpha}\sqrt{2\sigma^2(\theta_0)}}{\sigma(\theta_1 + \theta_0)}\right] \tag{48}$$

If $\omega < 1$, Eq. (48) becomes:

$$\beta_{\omega<1} = err\left[\frac{\theta_1 - Q_{1-\alpha}\sqrt{2\sigma^2(\omega\theta_0)}}{\sigma(\theta_1 + \omega\theta_0)}\right] > \beta_{\omega=1} \tag{49}$$

This effect, so-called "shadow effect," induces a significant loss in detection capability ($\beta\nearrow$) even for $\omega \approx 1$.

Many works have been done in order to restore the baseline ($\omega \to 1$) and all of them use a database recorded when a representative sample of void vehicle in passing through the RPM [22, 23]. An alternative method based on time series analysis has been developed to restore the baseline without using any prior knowledge about the vehicle and the experimental conditions hoping for gain in flexibility [24]. The latter is described below.

In the first place, the minimization of DT requires the implementation of an efficient smoothing filter, minimizing the high-frequency variance $\sigma^2(\theta_0)$ and subsequently the β risk, while preserving the temporal shape of the signal of interest $\theta_1(t_i)$. Thus, the nonlinear filter CST (Eqs. (36) and (37)) has proven efficient for this purpose. The single-channel RPM estimates the random variable $\rho_i\Delta t \sim \mathcal{P}(\theta_T\Delta t)$ at each time t_i such as:

$$\left[\widehat{\rho}_i^{**};\sigma\left(\widehat{\rho}_i^{**}\right)\right] = CST\left(\widehat{\rho}_i\right) \tag{50}$$

Estimations are continuously recorded into an historical memory with a depth m allowing calculating, at each time step t_i, the filtered logarithmic derivative $\hat{\dot{\rho}}_i$ of the signal:

$$\forall i \in [1, m],$$

$$\hat{\dot{\rho}}_i = (1 - \alpha_1)\frac{\widehat{\rho}_i^{**} - \widehat{\rho}_{i-l}^{**}}{\widehat{\rho}_i^{**}} - \alpha_1\hat{\dot{\rho}}_{i-1} \tag{51}$$

with α_1 and l being, respectively, a smoothing parameter and the derivative depth.

The trend of the signal, which can be constant, decreasing or increasing, is represented by a slope state D_i with values between -1 and 1 such as:

$$\hat{\rho}_i < -\alpha_2 \Rightarrow D_i = -1 \tag{52}$$

$$\left|\hat{\rho}_i\right| \le \alpha_2 \Rightarrow D_i = 0 \tag{53}$$

$$\hat{\rho}_i > \alpha_2 \Rightarrow D_i = 1 \tag{54}$$

with $\alpha_2 > 0$ being a parameter for variation significant.

The state of the signal S_i is labeled by a number between 1 and 8, defined as illustrated in **Figure 9**. The first line describes the passage of a vehicle containing a source without shadow effect; the second line corresponds to the passage of a dense vehicle with no source; and the third one to the passage of a dense vehicle containing a radioactive source (shadow shielding).

States S_i can be determined with knowledge of D_i and S_{i-1} using a sequential logic algorithm detailed under the form of a state diagram in **Figure 10**. To solve the problem, states 3 and 8 automatically pass to state 1 after a preset watchdog time τ_w.

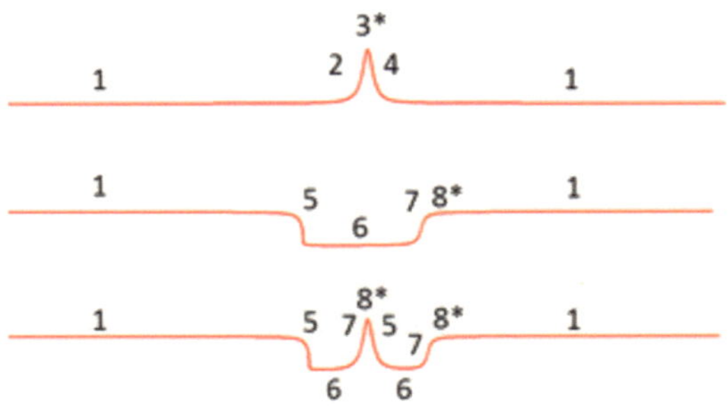

Figure 9. Schematic view of possible states of the system.

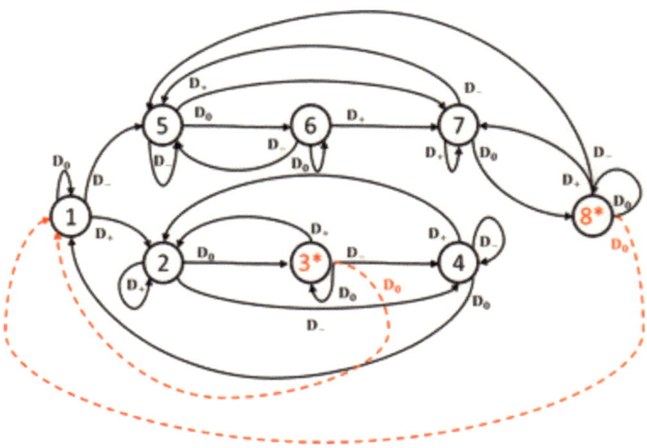

Figure 10. State diagram of the state determination algorithm.

http://dx.doi.org/10.5772/intechopen.71233

Knowing the state of the system, the baseline of the signal can be restored. The upper level (UL) (state 1) and the lower level LL (state 6) are firstly estimated in a recorded time series at time t_i:

$$z_{UL} = \underset{1 \le k \le \xi}{\arg} (S_{i-k} = 1) \tag{55}$$

$$z_{LL} = \underset{1 \le k \le \xi}{\arg} (S_{i-k} = 6) \tag{56}$$

$$UL = \frac{1}{\dim(Z_{UL})} \sum_{k=1}^{\dim(Z_{UL})} \widehat{\rho}_{i-k}^{**} \tag{57}$$

$$LL = \frac{1}{\dim(Z_{LL})} \sum_{k=1}^{\dim(Z_{LL})} \widehat{\rho}_{i-k}^{**} \tag{58}$$

The baseline is restored to obtain corrected count rate estimations $\widehat{\rho}^{***}$ such as:

$\forall k \in [\![1; \xi]\!]$,

$$S_{i-k} \in \{1; 2; 3; 4\}, \widehat{\rho}_{i-k}^{***} = \widehat{\rho}_{i-k}^{**} \tag{59}$$

$$S_{i-k} \in \{5; 6; 7; 8\}, \widehat{\rho}_{i-k}^{***} = \widehat{\rho}_{i-k}^{**} + UL - LL \tag{60}$$

Figure 11 illustrates the baseline restoration: the correction algorithm enables the detection of a source originally hidden by shadow effect. A simulation study has shown the significant gain in detection probability with the maintaining of a stable false detection rate [24].

The conception of a RPM primarily consists in designing detection blocks with a maximized sensitivity according to the application view and cost-effectiveness strategies. Signal processing is then to be implemented in the system in order to tune its detection capabilities.

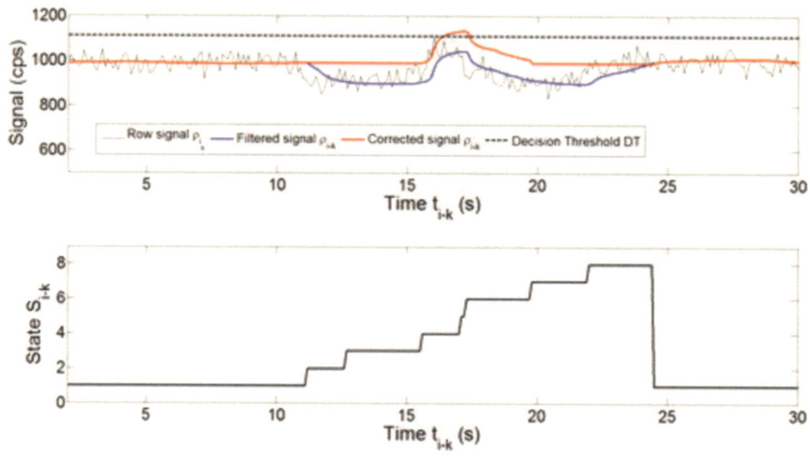

Figure 11. Signal and state evolutions over the simulation of a source passing into a dense vehicle.

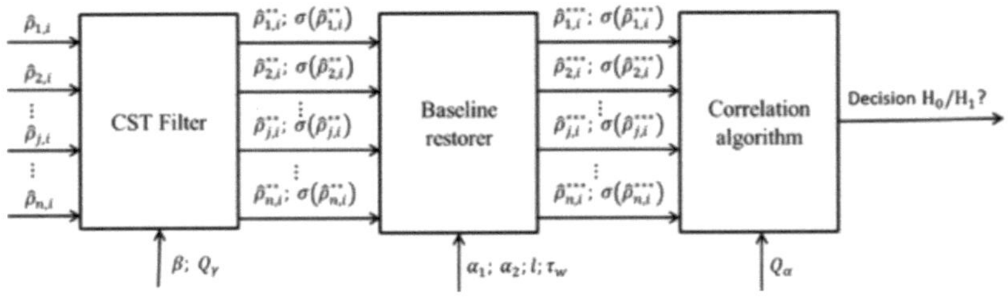

Figure 12. Schematic of a system based on correlation detection.

The improvement of RPM performance forms an active topic of research. It has notably been established that the spectral analysis of the signal, even for unresolved detectors, allows a gain in detection performance [25]. Another upgrade can be achieved by time series analysis techniques, especially when RPM are deployed in a network, which allows the implementation of correlation methods [26]. **Figure 12** presents the schematic of a RPM network implementing n channels and dedicated to moving source detection.

The network configuration enables two complementary types of detection: the first one based on traditional temporal analysis of individual channel $H_1|\widehat{\rho}_1, \ldots, \widehat{\rho}_n$, the second one based on frequency analysis, searching for a phase ϕ maximizing the correlation between channels $H_1|\varphi_{\widehat{\rho}_1, \ldots, \widehat{\rho}_n}$. When the network is linear and the source carrier has a constant velocity, this phase corresponds to the periodic echo of the signal increase on the first channel as seen on the other channels. The difference in nature between both methods introduces a quantitative information gain, and thus a potential improvement in detection capability [27–29].

A correlation vector R_φ is calculated by scanning the product of all channels with phase φ such as:

$\forall \varphi \in \left[1; \left\lfloor \frac{\xi-1}{n-1} \right\rfloor\right]$,

$$R_\varphi = \widehat{\rho}^{***}_{1,i-1} \widehat{\rho}^{***}_{2,i-(\varphi+1)} \widehat{\rho}^{***}_{3,i-(2\varphi+1)} \cdots \widehat{\rho}^{***}_{j,i-[(j-1)\varphi+1]} \cdots \widehat{\rho}^{***}_{n,i-[(n-1)\varphi+1]}$$
$$R_\varphi = \prod_{j=1}^{n} \widehat{\rho}^{***}_{j,i-[(j-1)\varphi+1]} \tag{61}$$

For the vehicle passing from detectors 1 to n.

The algorithm firstly determines a phase φ_0 maximizing R_φ, then the significance of the associated temporal correlation is evaluated with a hypothesis test in which $\mathbf{H_0}$ is the null hypothesis (no echo detected) and $\mathbf{H_1}$ is the detection hypothesis (echo detected). Values of R_φ are compared to the mean and variance of their distribution. Mean $\overline{R}$ and empirical variance $\sigma^2(R_\varphi)$ are calculated according to:

$$\overline{R} = \left\lfloor \frac{n-1}{\xi-1} \right\rfloor \sum_{\varphi=1}^{\left\lfloor \frac{\xi-1}{n-1} \right\rfloor} R_\varphi \tag{62}$$

$$\sigma^2(R_\phi) = \left\lfloor \frac{n-1}{\xi-1} \right\rfloor \sum_{\varphi=1}^{\left\lfloor \frac{\xi-1}{n-1} \right\rfloor} \left(R_\varphi - \overline{R}\right)^2 \tag{63}$$

The detection test reads:

Algorithm 2:

If $R_{\phi_0} > \overline{R} + Q_{1-\alpha}\sqrt{2\sigma^2(R_\phi)}$

Then $\mathbf{H_1}$ is accepted

Else $\mathbf{H_0}$ is accepted

The use of the empirical variance $\sigma^2(R_\phi)$ ensures a significant gain in detection capability under challenging signal-to-noise ratios [27, 28]. However, an algorithmic refinement is achieved with the introduction of a Normal law *a priori* on count rate distributions [29]. Thus, a modified variance $\sigma^2(R_\phi)$ is obtained by the estimation of individual variances $\sigma^2(\rho_{j,i})$ provides by the nonlinear filter (cf. Eq. (37)) for every detector j and memory slot i. Its calculation is presented in the following recursive formula:

$$\forall k \in [2; n] \,\&\, \forall \varphi \in \left[\!\left[1; \left\lfloor \frac{\xi-1}{n-1} \right\rfloor \right]\!\right],$$

$$\sigma^2(R_\phi)\big|_k = \sigma^2(R_\phi)\big|_{k-1}\left[\sigma^2\left(\widehat{\rho}^{**}_{k,i-(k-1)\phi+1}\right) + \left(\widehat{\rho}^{**}_{k,i-(k-1)\phi+1}\right)^2\right] + \sigma^2\left(\widehat{\rho}^{**}_{k,i-(k-1)\phi+1}\right)\left(R_\phi\big|_{k-1}\right)^2 \tag{64}$$

with

$$R_\varphi\big|_{k-1} = \widehat{\rho}^{**}_{1,1} \tag{65}$$

$$\sigma^2(R_\varphi)\big|_1 = \sigma^2\left(\widehat{\rho}^{**}_{1,i-1}\right) \tag{66}$$

The detection algorithm mixes the detection according to each individual channel with the detection using the correlation factor. In both cases, a decision threshold (*DT*) is calculated as a function of a false detection risk α. Let $DT_{j,\phi}$ the decision threshold associated with the channel j and the phase ϕ reads:

$$\forall j \in [1; n] \,\&\, \forall \varphi \in \left[\!\left[1; \left\lfloor \frac{\xi-1}{n-1} \right\rfloor \right]\!\right],$$

$$DT_{j,\varphi} = Q_{1-\alpha}\sqrt{2\sigma^2\left(\widehat{\rho}^{**}_{j,i-(j-1)\varphi+1}\right)} \tag{67}$$

And let DT_{R_ϕ} be the decision threshold associated with the correlation factor R_ϕ such as

$$\forall \varphi \in \left[\!\left[1; \left\lfloor \frac{\xi - 1}{n - 1} \right\rfloor \right]\!\right],$$

$$DT_{R_\phi} = Q_{1-\alpha}\sqrt{2\sigma^2(R_\varphi)\big|_n} \tag{68}$$

Algorithm 2 presents the mechanism of the cumulative detection.

Algorithm 3:

If $\forall j \in [1; n] \& \forall \varphi \in \left[\!\left[1; \left\lfloor \frac{\xi-1}{n-1} \right\rfloor \right]\!\right], DT_{j,\varphi} - \widehat{\rho}^{**}_{j,i-(j-1)ö+1} \leq 0,$

And if $\forall \varphi \in \left[\!\left[1; \left\lfloor \frac{\xi-1}{n-1} \right\rfloor \right]\!\right],\ DT_{R_\varphi} - R_\varphi \leq 0,$

Then, the detection hypothesis H_0 is accepted and the hypothesis H_1 is rejected,

Else if $\exists \phi \in \left[\!\left[1; \left\lfloor \frac{\xi-1}{n-1} \right\rfloor \right]\!\right] \& \exists j' \in [1; n], DT_{j,\varphi} - \widehat{\rho}^{**}_{j,i-(j-1)\varphi+1} > 0,$

Then, the detection hypothesis H_1 is accepted and the hypothesis H_0 is rejected,

Or if $\exists \varphi' \in \left[\!\left[1; \left\lfloor \frac{\xi-1}{n-1} \right\rfloor \right]\!\right], DT_{R_{\varphi'}} - R_{\varphi'} > 0,$

Then, the detection hypothesis H_1 is accepted and the hypothesis H_0 is rejected,

And, the velocity of the source is equal to $\frac{L}{\varphi' \Delta t}$ where L is the distance between detectors.

It has been proven in [27–29], the largely significant added-value in term of detection capability permits by the implementation of the correlation based detection. The true detection rate is increased while maintaining very low false alarm rate. **Figure 13** presents a system realized by the CEA which implements the correlation method [30].

All of these algorithms will be implemented in a dedicated DSP card [3] and the compliance of the RPM system will the standard ANSI42-35 will be tested in due course [31].

Figure 13. Photography of the RPM prototype (Katrina) developed by the CEA in the framework of the SECUR-ED project funded by the European Commission [30].

6. Conclusion

Different count rate processing methods have been presented in this chapter: an adaptive smoother, a background discrimination method and two algorithms improving the detection of moving sources. In these algorithms, frequentist inferences are realized on the basis of measured data. These types of approaches are well suited for real-time processing, allowing taking decision with very few iterations, compared to Bayesian inferences which are more suited for post-processing analyses.

The nonlinear smoother is proved to be a key building block in radiation monitors, delivering a fine estimation of count rate expectation with a minimized associated variance. Both expectation and variance estimations are used to apply hypothesis tests addressing many problematics in radiation monitoring such as for instance those already developed hereby: compensation and RPM network.

Author details

Romain Coulon* and Jonathan Dumazert

*Address all correspondence to: romain.coulon@cea.fr

French Alternative Energies and Atomic Energy Commission, Gif-sur-Yvette, France

References

[1] Knoll GF. Radiation Detection and Measurement. 4th ed. Hoboken: Wiley; 2010. p. 625-704 & 66-85

[2] Jordanov VT, Knoll GF, Huber AC, Pantazis JA. Digital techniques for real-time pulse shaping in radiation measurements. Nuclear Instruments and Methods A. 1994;**353**:261-264

[3] French M, Thevenin M, Hamel M, Montbarbon E. A histogram-difference method (HDM) for neutron/gamma discrimination using liquid and plastic scintillators. IEEE Transactions on Nuclear Science. 2017;**68**(8):2423-2432

[4] Rohée E, Coulon R, Carrel F, Dautremer T, Barat E, Montagu T, et al. Benchmark of the non-parametric Bayesian deconvolution method implemented in the SINBAD code for X/γ rays spectra processing. Nuclear Instruments and Methods A. 2016;**836**:91-97

[5] Timmermann KE, Nowak RD. Multiscale Bayesian estimation of Poisson intensities. In: Conference Record of the Thirty-First Asilomar Conference on Signals, Systems and Computers, 1997:85-90

[6] Rodriguez G. Lecture Notes on Generalized Linear Models. 2007. URL: http://data.princeton.edu/wws509/notes/

[7] Londen-Thurgood RM, Pople J. A new ratemeter concept for radiation measuring instruments. Nuclear Instruments and Methods. 1981;**184**(2–3):533-536

[8] Savic Z. Some software algorithms for microprocessor ratemeters. Nuclear Instruments and Methods A. 1991;**301**(3):517-522

[9] Basseville M, Nikiforov IV. Detection of Abrupt Changes: Theory and Application. Prentice-Hall, Upper Saddle River; 1993

[10] Fehlau PE. Comparing a recursive digital filter with the moving-average and sequential probability-ratio detection methods for SNM portal monitors. IEEE Transactions on Nuclear Science. 1993;**40**(2):143-146

[11] Apostolopoulos A. On-line statistical processing of radiation detector pulse trains with time-varying count rates. Nuclear Instruments and Methods A. 2008;**595**:464-473

[12] Coulon R, Dumazert J, Kondrasovs V, Normand S. Implementation of a nonlinear filter for online nuclear counting. Radiation Measurements. 2016;**87**:13-23

[13] Gilmore G. Practical Gamma-ray Spectrometry. 2nd ed. Hoboken: Wiley; 2008. p. 114-119

[14] Coulon R, Dumazert J, Kondrasovs V. Estimation of nuclear counting by nonlinear filter based on a hypothesis test and a double exponential smoothing. IEEE Transactions on Nuclear Science. 2016;**63**(5):2671-2676

[15] Boudergui K, Frelin A-M, Kondrasovs V, Normand S. Integrated sensor handled by robot for dose rate measurement. In: Information System on Occupational Exposure, 2010:1-5

[16] Iwanowska J, Swiderski L, Moszynski M. Neutron/gamma discrimination properties of composite scintillation detectors. Journal of Instrumentation. 2011;**6**:P07007

[17] Mori C, Kumanomido H, Watanabe T. Background compensated GM counter for the measurement of low level β-activity. Nuclear Instruments and Methods. 1983;**211**(2–3):429-432

[18] Mori C, Kumanomido H, Watanabe T. Application of a background-compensated Geiger–Müller counter to a survey meter. Nuclear Instruments and Methods A. 1984;**228**(1):177-182

[19] Coulon R, Montagu T, Schoepff V. Reliability improvement for anisotropic biased compensation \alpha/\beta contamination meter. Nuclear Instruments and Methods A. 2016;**837**:44-49

[20] Dumazert J, Coulon R, Bertrand GHV. Compensated bismuth-loaded plastic scintillators for neutron detection using low-energy pseudo-spectroscopy. Nuclear Instruments and Methods A. 2016;**819**:25-32

[21] Dumazert J, Coulon R, Kondrasovs V. Compensation scheme for online neutron detection using a Gd-covered CdZnTe sensor. Nuclear Instruments and Methods A. 2017;**857**:7-15

[22] Lo Presti CA, Weier DR, Kouzes RT, Schweppe JE. Baseline suppression of vehicle portal monitor gamma count profiles: A characterization study. Nuclear Instruments and Methods A. 2006;**562**(1):281-197

[23] Burr T, Myers K. Effects of background suppression of gamma counts on signal estimation. Applied Radiation and Isotopes. 2009;**67**(9):1729-1737

[24] Coulon R, Dumazert J. Shadow shielding compensation for moving sources detection. IEEE Transactions on Nuclear Science. 2017;**64**(7):1641-1646

[25] Runkle RC. Analysis of spectroscopic radiation portal monitor data using principal components analysis. IEEE Transactions on Nulcear Science. 2006;**53**(3):1418-1423

[26] Robinson SM, Bender SE. Time series evaluation of radiation portal monitor data for point source detection. IEEE Transactions on Nuclear Science. 2009;**56**(6):3688-3693

[27] Coulon R, Kondrasovs V, Boudergui K, Normand S. Moving sources detection system. In: ANIMMA; 2013:1-4

[28] Coulon R, Kondrasovs V. Moving sources detection algorithm for radiation portal monitors used in a linear network. IEEE Transactions on Nuclear Science. 2014;**61**(4):2189-2194

[29] Dumazert J, Coulon R, Kondrasovs V. A robust hypothesis test for the sensitive detection of constant speed radiation moving sources. Nuclear Instruments and Methods A. 2015;**795**:335-342

[30] Boudergui K, Kondrasovs V, Coulon R. New monitoring system to detect a radiative material in motion. In: Proceedings of ANIMMA; 2013:1-5

[31] IEEE Standard Association, editors. American National Standard for Evaluation and Performance of Radiation Detection Portal Monitors for Use in Homeland Security (ANSI N42.35). Washington: ANSI; 2016